Guidance for Assessing Exemption Requests from the Nuclear Power Plant Licensed Operator Staffing Requirements Specified in 10 CFR 50.54(m)

I0482744

Final Report

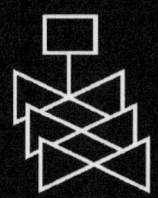

U.S. Nuclear Regulatory Commission
Office of Nuclear Regulatory Research
Washington, DC 20555-0001

NUREG-1791

Guidance for Assessing Exemption Requests from the Nuclear Power Plant Licensed Operator Staffing Requirements Specified in 10 CFR 50.54(m)

Final Report

Manuscript Completed: July 2005
Date Published: July 2005

Prepared by
J. Persensky, A. Szabo/U.S. Nuclear Regulatory Commission
C. Plott, T. Engh/Micro Analysis & Design
V. Barnes/Performance, Safety, & Health Associates

A. Szabo, NRC Project Manager
J. Persensky, Senior Technical Advisor

Prepared for
Division of Risk Analysis and Applications
Office of Nuclear Regulatory Research
U.S. Nuclear Regulatory Commission
Washington, DC 20555-0001

ABSTRACT

Regulations set forth by the U.S. Nuclear Regulatory Commission (NRC) prescribe the qualifications and staffing levels for licensed operating personnel for nuclear power plants. The introduction of advanced reactor designs and increased use of advanced automation technologies in existing nuclear power plants will likely change the roles, responsibilities, composition, and size of the crews required to control plant operations. Current regulations regarding control room staffing, which are based upon the concept of operation for existing light-water reactors, may no longer apply. Therefore, applicants for an operating license for an advanced reactor, and current licensees who have implemented significant changes to existing control rooms, may submit applications for exemptions from current staffing regulations. The NRC staff is responsible for reviewing the exemption requests and must determine whether the staffing proposals provide adequate assurance that public health and safety will be maintained at a level that is comparable to that afforded by compliance with the current regulations.

This guidance provides a process for systematically reviewing and assessing these submittals. It details the information, data, and review criteria necessary to review the exemption request. The information and data from an exemption request could include (1) the description of the request, the concept of operations, and operational conditions considered, (2) supporting analyses and documentation from operating experience, functional requirements analysis and function allocation, task analysis, job definition, and staffing plan, and (3) data and analysis from validation exercises performed to demonstrate the effectiveness and safety of the proposed staffing plan.

The information collections contained in this guidance are covered by the requirements of 10 CFR Part 50, which were approved by the Office of Management and Budget, approval number 3150-0011.

Public Protection Notification

FOREWORD

This document provides guidance for the NRC staff to systematically review and assess requests by licensees of nuclear power plants for exemption from the licensed operator staffing requirements of Title 10, Part 50 of the Code of Federal Regulations (10 CFR 50) contained in 10 CFR 50.54(m). The purpose of the NRC's review is to ensure public health and safety by verifying that the applicant's staffing plan and supporting analyses sufficiently justify the requested exemption.

The increased use of advanced automation technologies in existing nuclear power plants and the introduction of advanced reactor designs will likely change the roles, responsibilities, composition, and size of the crews required to control plant operations. Current regulations regarding control room staffing, which are based on the concept of operation for existing light-water reactors, may no longer apply. Licensees of nuclear power plants who have implemented significant changes to existing control rooms or who have introduced increased use of advanced automation technologies may submit applications for exemption from the requirements. Likewise, because of the anticipated changes in operator roles and responsibilities in new reactor designs, an applicant for an operating license for a new reactor may wish to seek exemption from the current licensed operator staffing requirements.

This document identifies the type of information and data that should be considered in a regulatory review of an exemption request which includes: (1) the description of the request, the concept of operations, and operational conditions considered, (2) supporting analyses and documentation from operating experience, functional requirements analysis and function allocation, task analysis, job definition, and staffing plan, and (3) data and analysis from validation exercises performed to demonstrate the effectiveness and safety of the proposed staffing plan. The review process described in this document addresses all of these elements and includes guidance for each review step.

In reviewing exemption requests, the NRC staff must determine whether the licensees' staffing proposals provide adequate assurance that public health and safety will be maintained at a level that is comparable to compliance with the current regulations. This guidance provides a strong technical basis to inform the decision-making process and allow the staff to make good technical judgments. This approach is consistent with the performance goals of the NRC's Strategic Plan for Fiscal Years 2004–2009 (NUREG-1614, Vol. 3, as well as Chapter 18.0 of the NRC's Standard Review Plan (NUREG-0800) and Rev. 2 of NUREG-0711, "Human Factors Engineering Program Review Model." It is also consistent with NUREG/IA-0137, "A Study of Control Room Staffing Levels for Advanced Reactors."

Carl J. Paperiello, Director
Office of Nuclear Regulatory Research
U.S. Nuclear Regulatory Commission

CONTENTS

Figures

Tables

EXECUTIVE SUMMARY

U.S. Nuclear Regulatory Commission (NRC) regulations prescribe the qualifications and staffing levels for licensed operating personnel for nuclear power plants. The design features and concepts of operation for new generations of advanced reactors, as well as the increased use of advanced, automated, and digital systems in existing plants, may lead applicants to request variations in the prescribed number, composition, or qualifications of licensed personnel. This will require applicants to submit exemption requests from applicable regulations included primarily in Title 10, Section 50.54(m), of the *Code of Federal Regulations* (10 CFR 50.54(m)).

Staff of the NRC will review these exemption requests to determine their acceptability. The purpose of this review is to ensure public health and safety by verifying that the applicant's staffing plan and supporting analyses sufficiently justify the requested exemption. The applicant's submittal should include (1) the description of the request, the concept of operations, and operational conditions considered, (2) supporting analyses and documentation from the operating experience, functional requirement analysis and function allocation, task analysis, job definition, and staffing plan, and (3) data and analysis from validation exercises performed to demonstrate the effectiveness and safety of the proposed staffing plan.

This document describes a review process that addresses all of these elements and includes the following guidance for each review step:

- a *discussion* of the review step and why it needs to be addressed
- *data and information required* to support the review step
- *review criteria* for evaluating the submittals
- *additional information* that may be useful in performing the review

The guidance also includes, in Appendix A, a series of checklists for each review step. Appendix B provides a glossary of terms used in the guidance, and Appendix C offers a list of references.

The review steps are as follows:

(1) Review the Exemption Request

The NRC staff conducts a general review of the requested exemption(s) to determine the scope of the request(s). As part of this review, the staff also identifies any new or modified concepts, or changes in meanings for terms included in the regulations (e.g., operator, control room, unit) introduced as part of the application.

(2) Review the Concept of Operations

The purpose of reviewing the concept of operations is to gain an understanding of the role of control personnel in overall plant operations. Understanding the applicant's intended concept of operations also establishes the context for subsequent steps in the review.

(3) Review the Operational Conditions

The purpose of the review of the operational conditions is to ensure that the applicant analyzed the operational conditions which present the greatest potential challenges to the effective and safe performance of control personnel, under the conditions of the requested exemption, and that the analysis supports the exemption request. The staff evaluates the operational conditions defined by the applicant for completeness and uses them with the other review steps to assess the exemption request.

(4) Review Operating Experience

The staff reviews the operating experience to ensure that the applicant has performed a review of relevant operational experience to identify and address staffing-related lessons learned that may be important to the exemption request. This review may include plants with similar designs, plants that have implemented similar technologies, or similar concepts of operation from other industries.

(5) Review the Functional Requirements Analysis and Function Allocation

This review has two purposes. The first purpose is to ensure that the applicant has defined and evaluated the impact of the exemption request on the plant/system functions that must be performed to satisfy plant safety objectives. The second purpose is to ensure that the allocation of functions to humans and systems has resulted in a role for personnel that uses human strengths, avoids human limitations, and can be performed under the operational conditions evaluated in the exemption request.

(6) Review the Task Analysis

The purpose of the task analysis review is to ensure that the applicant's analysis identifies the specific tasks that are needed to accomplish functions and their staffing implications. For each task, the applicant should address the information, control, and task-support requirements in its task analysis, as applicable. The reviewer will ensure that the applicant identified any issues related to task timing, workload, situation awareness, and resource conflicts that would affect staffing assignments.

(7) Review the Job Definitions

The purpose of the job definition review is to confirm that the applicant has established clear and rational job definitions for the personnel who will be responsible for controlling the plant, in the case of a new plant design. For an existing plant in which new systems will be implemented, the purpose of the review is to ensure that the applicant has retained clear and rational job definitions for control room personnel.

(8) Review the Staffing Plan

The purpose of the staffing plan review is to ensure that the applicant has systematically analyzed the requirements for the numbers of qualified personnel that are necessary to operate the plant safely under the operational conditions analyzed.

(9) <u>Review of Additional Data and Analyses</u>

The reviewer may find that additional data and types of reviews may be needed to complete the review of the exemption request(s). The reviewer could require data from areas such as human reliability analysis; human-system integration; knowledge, skills, and abilities; procedures; and training analysis. Data submitted by the applicant for these review areas would provide further justification in support of the exemption request.

(10) <u>Review the Staffing Plan Validation</u>

The most important step is the review of the applicant's staffing plan validation. The purpose of reviewing the validation of the staffing plan is to ensure that the applicant fully considered the dynamic interactions between the plant design, its systems, and control personnel for the operational conditions identified for the exemption request.

(11) <u>Determine Acceptability of the Exemption Request</u>

In this step, the NRC staff makes a final decision regarding the acceptability of the exemption request. The decision will be based on the aggregate findings from the previous steps of the review.

ACKNOWLEDGMENTS

We would like to acknowledge the following people who provided valuable feedback during the early development of this document as members of our peer review committee:

Joseph DeBor	Consultant
Richard Eckenrode	Office of Nuclear Reactor Regulation, NRC (retired)
Robert B. Fuld	Westinghouse Nuclear Automation
Greg Galletti	Office of Nuclear Reactor Regulation, NRC
Lew Hanes	Electric Power Research Institute
John Hawley	Army Research Laboratory
Jacques Hugo	Pebble Bed Modular Reactor
Scott Malcolm	Atomic Energy of Canada Limited
John O'Hara	Brookhaven National Laboratory
Kevin Williams	Office of Nuclear Reactor Regulation, NRC

We would also like to acknowledge the following people who provided valuable feedback during the internal review of this document:

Matthew Chiramal	Office of Nuclear Reactor Regulation, NRC
James Bongarra	Office of Nuclear Reactor Regulation, NRC

ABBREVIATIONS

CFR   *Code of Federal Regulations*

HFE   human factors engineering

HSI   human systems interface

KSA   knowledge, skills, and abilities/aptitudes

NRC   U.S. Nuclear Regulatory Commission

NUREG   NRC technical report designation (<u>Nu</u>clear <u>Reg</u>ulatory Commission)

PRA   probabilistic risk assessment

RAI   request for additional information

SA   situation awareness

PART I
INTRODUCTION

1. OVERVIEW OF THE GUIDANCE DOCUMENT

1.1   Purpose and Scope

The purpose of this document is to provide the staff of the U.S. Nuclear Regulatory Commission (NRC) with a process for evaluating requests for exemption from one or more of the requirements specified in Title 10, Section 50.54(m) of the *Code of Federal Regulations* (10 CFR 50.54(m)). Other regulations may be impacted by a request for exemption from the current staffing regulations. For example, a request for exemption may be based on the implementation of advanced technologies that change the roles and responsibilities of personnel licensed under 10 CFR Part 55. Because the nature of potential exemption requests based on advanced technologies is currently unknown, this guidance provides a flexible review process and a set of systematic methods that the NRC can use to evaluate a wide range of staffing-related exemption requests.

1.2   Background

The introduction of advanced reactor designs and the increased use of advanced automation in existing nuclear power plants will likely change the roles, responsibilities, composition, and size of the crews required to control plant operations. The design features and concepts of operation for new generations of advanced reactors, as well as the introduction of new automated or digital systems into existing plants, may lead to reductions in staff size and a changing role for the operator. For the purposes of this guidance document, the term, *concept of operation*, refers to a description of how a licensee's or applicant's organizational structure, staffing, and management framework relate to the systems, design, and operational characteristics of the plant. Current regulations regarding control room staffing, which are based upon the concept of operation for existing light-water reactors, may no longer apply. Therefore, applicants for an operating license for an advanced reactor, and current licensees who have implemented significant changes to existing control rooms, may submit applications for exemptions from current staffing regulations. The NRC staff will review the exemption requests and determine whether the staffing proposals will provide adequate assurance that public health and safety will be maintained at a level that is comparable to compliance with the current regulations.

1.3   Impact of New Technologies on the Roles and Responsibilities of Licensed Personnel

There is a relatively wide range of potential changes to reactor designs and to the technologies that will be available for maintaining operational control of new and existing nuclear power plants. Simplified designs and operations, increased use of advanced automation, and new technologies for human-system interfaces (HSIs) will change the role of the human in plant operations.

Many advanced reactor designs incorporate passive safety features that require minimal operator intervention to mitigate accidents in the event of malfunction. These passive safety features are based on natural forces, such as convection and gravity, making safety functions less dependent on active systems and components, such as pumps and valves. The passive features allow operators more time to perform safety actions. The increased time available to respond to events could reduce the number of personnel required on each shift, because there would be time to augment the control staff to respond to the event.

For both advanced plant control room designs and some potential upgrades to existing control rooms, the increased use of automated control, monitoring, and protection systems will bring the plant back to normal conditions or to a safe shutdown state without the need for operator action. In addition, this reduced need for human intervention may, in turn, reduce the number of people required to control plant operations.

Automated systems that support and supplement human cognitive functions associated with controlling the plant may also affect staffing configurations in both advanced and existing plant control rooms. Technological advances, such as intelligent agents, computer-supported cooperative work, knowledge engineering, and knowledge-based systems, are leading to designs of automated systems that enhance control room personnel capabilities for monitoring, disturbance detection, situation assessment, response planning, and response execution.

Along with these advances in automation technology, new HSI technologies are also emerging. Rather than having control rooms with panels full of controls and displays, there will be "control suites," consisting of a set of computer displays and input devices. Information can be displayed dynamically across the monitors in the display and enhanced auditory signals, such as speech, will be possible. An array of input capabilities including touch, gesture, and speech are also possibilities. Intelligent support systems can enhance the timing and form of information provided to operations personnel and the management of both automated and manual actions. These advances may reduce the number of plant personnel required to maintain operational control of a single reactor or may allow plant personnel to maintain control of multiple units from one control suite.

In addition, advances in the bandwidth and reliability of telecommunications technologies (including wireless) create the possibility of remote operations from both remote control suites and from portable devices, such as laptop computers or personal digital assistants. Telecommunications technologies may allow personnel to monitor and control multiple reactors from remote locations, though security constraints may limit the use of these technologies.

Implementation of advanced technologies may change some or all of the HSI elements upon which current staffing approaches are based. The advanced technologies will likely result in changes to the allocation of functions and tasks among personnel and systems. The character of the functions and tasks may also change, resulting in changes to the number and qualifications of personnel needed. For example, because monitoring of plant parameters and most control tasks may be fully automated, on-site operations personnel may be able to perform the majority of site maintenance tasks or other tasks that are currently assigned to other plant personnel.

Some of these potential changes to plant designs and to the systems used to control plant operations may make obsolete the concept of a traditional reactor control room staffed by a crew of licensed operators. To summarize, the possible changes to current control room staffing approaches include the following (among others):

- smaller control room crews than currently required

- smaller, or similarly sized, crews that are responsible for a greater number of reactors

- control suites that allow operational control of multiple reactors with the same set of controls and displays

- offsite operations of one or more reactors

- the introduction of new staff positions with new qualifications

1.4   Limitations of the Current Regulatory Structure

The current requirements for control room staffing are primarily contained in 10 CFR 50.54(m). For convenience, Table 1 presents the requirements of 10 CFR 50.54(m)(2)(i) which prescribe licensed operator staffing levels. Several limitations in the scope of these requirements, as well as the requirements of 10 CFR 50.54(m)(2)(ii), (iii), and (iv), exist. Some key assumptions are also implicit in the requirements. These limitations and assumptions include:

- there is a maximum of three units and three control rooms

- the number of control rooms does not exceed the number of units

- there are no more than two units per control room

- there is always at least one operator at the controls for each unit (10 CFR 50.54(m)(2)(iii))

- there is always at least one, and sometimes two additional operator(s) on site, for each unit in operation

- there is at least one senior operator on site at all times (10 CFR 50.54(m)(2)(ii))

- there is one senior operator in the control room for each unit in operation (10 CFR 50.54(m)(2)(iii))

- there is one more senior operator than the number of units operating when multiple units are in operation in more than one control room, except when three units are in operation in two control rooms

- operator and senior operator are the only two job functions addressed by the *Code of Federal Regulations*, and their roles, responsibilities, and qualifications are as defined in 10 CFR Part 55

Table 1. Minimum Requirements[1] Per Shift for On-Site Staffing of Nuclear Power Units by Operators and Senior Operators Licensed Under 10 CFR Part 55

Number of Nuclear Power Units Operating[2]	Position	One Unit	Two Units		Three Units	
		One Control Room	One Control Room	Two Control Rooms	Two Control Rooms	Three Control Rooms
None	Senior Operator	1	1	1	1	1
	Operator	1	2	2	3	3
One	Senior Operator	2	2	2	2	2
	Operator	2	3	3	4	4
Two	Senior Operator		2	3	3[3]	3
	Operator		3	4	5[3]	5
Three	Senior Operator				3	4
	Operator				5	6

[1] Temporary deviations from the numbers required by this table shall be in accordance with criteria established in the unit's technical specifications.

[2] For the purpose of this table, a nuclear power unit is considered to be operating when it is in a mode other than cold shutdown or refueling, as defined by the unit's technical specifications.

[3] The number of required licensed personnel when the operating nuclear power units are controlled from a common control room is two senior operators and four operators.

Finally, 10 CFR 50.54(m)(2)(iv) requires the following:

> Each licensee shall have present, during alteration of the core of a nuclear power unit (including fuel loading or transfer), a person holding a senior operator license or a senior operator license limited to fuel handling to directly supervise the activity and, during this time, the licensee shall not assign other duties to this person.

These assumptions and limitations reflect a concept of operation that is consistent with the design and operation of conventional light-water reactors. Also reflected is a "margin of safety" policy that suggests that there should be a sufficient number of operators and senior operators to safely operate the plant, plus one more, in case something happens to one of them.

1.5   Implications for the Review of Exemption Requests

Advanced reactor designs and the implementation of advanced technologies in existing plants will result in staffing configurations for operations that were not anticipated by the current regulations. As a result, licensees may request exemptions from current requirements as new technological innovations and concepts of operations are introduced. Licensees must submit these requests to the NRC for review.

When evaluating the exemption requests, reviewers will assess the impact of the staffing proposals on safety issues, such as the following:

- operators taking less active roles for ensuring the safety of the plant.

- operators having a greater range of roles and responsibilities in addition to their safety-related roles and responsibilities

- the need for operators to maintain situation awareness across a number of units and to potentially manage simultaneous operations across these units

- changes in the response times required from responsible personnel

- plant control capabilities provided by smaller, portable, or remote HSIs

- the interaction between the control room personnel and advanced HSIs, including intelligent support systems

- capabilities for managing and coordinating control functions among personnel who may be located remotely from each other

- changes in the qualifications of responsible personnel

- effective scheduling of a reduced number of personnel to optimize cognitive workload, minimize fatigue, and support situational awareness

In addition, concepts such as "the control room," "a unit," "at the controls," and "operator" may take on entirely new meanings that may require a new approach to licensing reviews regarding the personnel responsible for the safe operation of the plant.

1.6   Applicability

The guidance presented in this document is applicable when a request is submitted for an exemption from the requirements of 10 CFR 50.54(m). When an exemption is requested in accordance with 10 CFR 50.12, the applicant must submit evidence that the staffing proposal is adequate for the safe operation of the plant under all relevant operational conditions.

Within the overall regulatory framework, this guidance document presents a more detailed process for implementing the guidance contained in Section 6, "Staffing and Qualifications," of NUREG-0711, "Human Factors Engineering Program Review Model" (NRC, 2004). It is based on the guidance provided in Sections 13.1.2–13.1.3, "Operating Organization," and Chapter 18.0, "Human Factors Engineering," of NUREG-0800, "Standard Review Plan." (NRC, 2004). Figure 1 illustrates the relationship of this document to current regulations and other related guidance.

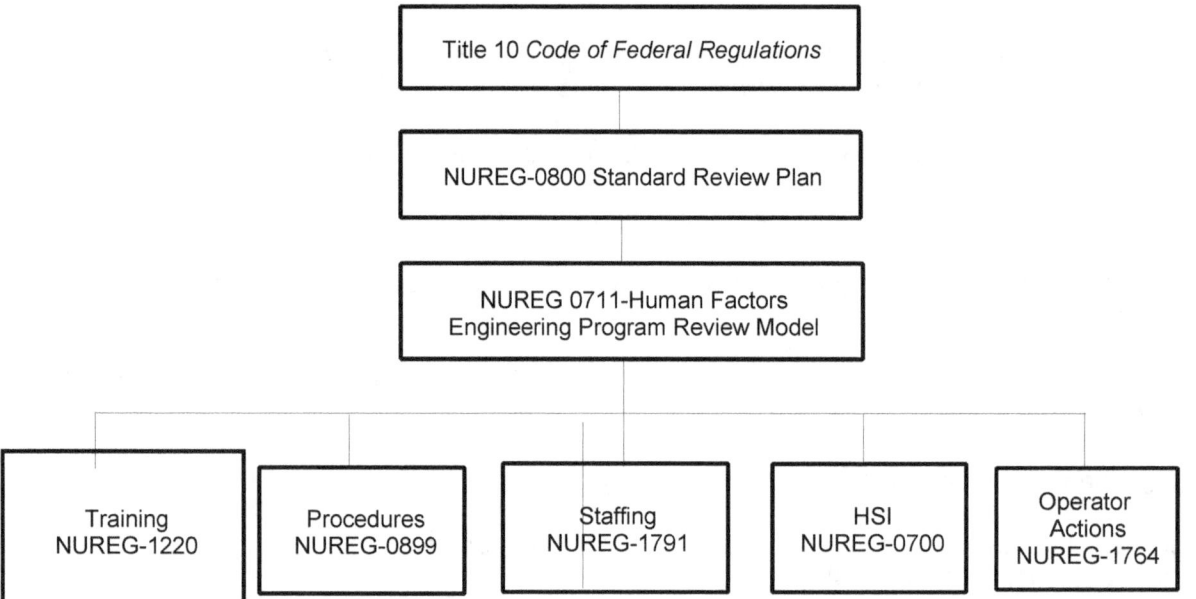

Figure 1. Relationship of this Document to Current Regulations and Other Guidance

1.7   Organization of the Guidance

In subsequent sections of this document, each step of the review process is presented in greater detail. For each review step, the guidance discusses background information on the step, defines important terms used in the review process, and presents criteria for conducting the review. When appropriate, the document discusses the types of data to be submitted by the applicant for each step and, at the end of each section, presents a listing of resources that may provide useful additional information to the reviewer. Appendix A includes checklists that may assist in organizing the reviewer's task. Appendix B provides a glossary of terms used in the guidance and Appendix C contains a list of references. If an applicant fails to provide sufficient information to perform the review, the staff should generate a request for additional information (RAI) to describe the needed data or information.

2. OVERVIEW OF THE REVIEW PROCESS

NUREG/CR-6838, "Technical Basis for Regulatory Guidance for Assessing Exemption Requests from the Nuclear Power Plant Licensed Operator Staffing Requirements Specified in 10 CFR 50.54(m)," provides methods, measures, criteria, and rationale for reviewing exemption requests, and comprises the basis for the guidance presented here. The process for reviewing staffing-related exemption requests consists of 11 steps. Figure 2 illustrates the overall flow of the review process.

The first step of the review process is to review the exemption request, which is a general review of the requested exemption(s), to determine the scope of the request(s). In addition, the staff identifies any new or modified concepts, or changes in meanings for terms included in the regulations (e.g., operator, control room, unit), introduced as part of the application.

The next step is to review the concept of operations to gain an understanding of the role of plant personnel in overall plant operations. Understanding the applicant's intended concept of operations also establishes the context for subsequent steps in the review.

The third step is to review the operational conditions considered by the applicant to justify the requested exemption(s). Of particular interest are those operational conditions that present the greatest challenges to the performance of licensed personnel working under the conditions included in the exemption. The staff evaluates the operational conditions defined by the applicant for completeness and uses them to assess the exemption request.

The next five steps focus on reviewing the required data and analyses from the submittals to verify that they are complete and provide adequate support for the exemption request. The following areas will be reviewed, as applicable:

- operating experience
- functional requirements analysis and function allocation
- task analysis
- job definitions
- staffing plan

The reviewer may find that additional data and types of reviews may be needed to complete the review of the exemption request(s). These additional reviews could require data from areas such as human reliability analysis, human-system integration, and knowledge, skills, and abilities (KSA) analysis. Data submitted for these review areas would provide further justification in support of the exemption request.

The most important step is the review of the applicant's staffing plan validation. Staffing plan validation refers to an evaluation using performance-based tests to determine whether the staffing plan meets performance requirements and acceptably supports safe operation of the plant. The final step in the review process is the final assessment of the exemption request to determine whether it is acceptable.

1. Review the Exemption Request

2. Review the Concept of Operations

3. Review the Operational Conditions

4. Review Operating Experience

5. Review the Functional Requirements Analysis and Function Allocation

6. Review the Task Analysis

7. Review the Job Definitions

8. Review the Staffing Plan

9. Review of Additional Data and Analyses

10. Review the Staffing Plan Validation

11. Determine Acceptability of the Exemption Request

Figure 2. The Exemption Request Review Process

PART II
EVALUATION OF EXEMPTION REQUESTS

1. REVIEW THE EXEMPTION REQUEST

1.1   Discussion

There are two reasons for performing an overall review of the exemption request. The first is to ensure that the reviewer understands the scope of the request. The second is to ensure that the applicant has submitted the necessary information to perform the review.

1.1.1   Scope of the Exemption Request

The applicant's request for exemption should be clear and specific about the portion(s) of Title 10, Section 50.54(m), of the *Code of Federal Regulations* (10 CFR 50.54(m)) from which an exemption is requested. The exemption request could include the following straightforward variations on the requirements of 10 CFR 50.54(m):

- a greater number of units controlled per control room

- a greater number of units for which an operator or senior operator is responsible

- changes in the responsibilities or qualifications of the control personnel, such as combining the responsibilities for operations and fuel handling

Applicants may also submit more complex requests, such as the following:

- the definition of new jobs that include functions not currently assigned to licensed operators

- control of operations at multiple sites from one control room

- an expanded definition of "at the controls" to include portable monitoring devices that would allow responsible personnel to monitor plant parameters and maintain operational control from either outside the control room or offsite during normal operations

These latter, nontraditional concepts of operations may result in the need to redefine terms such as "control room," "operator," and "at the controls." They may also result in the need for new operational terms and definitions. When this is the case, the applicant must indicate the need for, and provide definitions for, these new ideas and terms as part of the exemption request. When this type of information is not provided, or when the reviewer is uncertain of what is being requested, the staff should request clarification from the applicant.

An additional element of this portion of the review is to assess whether further exemptions from the regulations may be required. Given the number of possible types of exemptions that might be requested and the interrelationship of 10 CFR Part 50.54(m) to other NRC regulations, it is possible that exemptions from other regulations may also be required.

1.1.2 Information Completeness

The amount and level of detail required for the review will depend upon the nature of the exemption request. The reviewer may need a comprehensive physical representation (e.g., mockup, simulation) of the design of the proposed plant and its systems, as well as descriptions of plant responses to inputs and expected response times. The staff may also need detailed control room and human-system interface (HSI) representations. If only a few systems are being upgraded in an existing plant, the reviewer will require less extensive information, although detailed HSI representations for the new systems may be necessary. These representations may be needed to allow the reviewer to understand and evaluate the impact(s) of the requested exemption(s).

In addition, the reviewer should verify that the applicant's submittal meets the data requirements for the subsequent steps in the review process. Although the reviewer's information needs for subsequent steps may not be fully known during this initial step, by comparing the applicant's submittal against the data requirements for later steps in the review process, the reviewer may identify areas in which an RAI to the applicant will be necessary.

1.2   Applicant Submittals

The request for exemption should include the following elements:

- a description of the specific aspects of 10 CFR 50.54(m) from which an exemption is requested

- a physical representation of the plant and systems involved

- descriptions of plant/system responses to inputs and expected equipment response times

- a detailed representation of the control room, control suites, and/or the HSI to be used for monitoring and control actions

- definitions of any new terms used or definitions of terms whose meanings are changed

- information to meet the data requirements of subsequent review steps

1.3   Review Criteria

The reviewer should ensure that each of the following criteria has been met:

- Confirm that one or more exemptions to 10 CFR 50.54(m) is required.

- Confirm that exemptions from other, related regulations are either unnecessary or have been appropriately identified and described by the applicant. If additional exemptions are required that have not been identified by the applicant, the applicant should be informed and the review should be stopped until a complete request for exemptions is submitted.

- Confirm that the terms used in the submittal are fully defined.

- Confirm that adequate data and information have been submitted to meet the data requirements for the remainder of the review.

1.4   Additional Resources

The impact of the requested exemption on the following regulations and guidelines should be considered:

- 10 CFR 50.54(i), which states that, "except as provided in 55.13 of this chapter, the licensee may not permit the manipulation of the controls of any facility by anyone who is not a licensed operator or senior operator as provided in Part 55 of this chapter."

- 10 CFR 50.54(j), which states that, "apparatus and mechanisms other than controls, the operation of which may affect the reactivity or power level of a reactor shall be manipulated only with the knowledge and consent of an operator or senior operator licensed pursuant to Part 55 of this chapter present at the controls."

- 10 CFR 50.54(k), which states that, "an operator or senior operator licensed pursuant to part 55 of this chapter shall be present at the controls at all times during the operation of the facility."

- 10 CFR 50.54(l), which states that, "the licensee shall designate individuals to be responsible for directing the licensed activities of licensed operators. These individuals shall be licensed as senior operators pursuant to Part 55 of this chapter."

- 10 CFR 55.4, which provides definitions for controls, facility, licensee, operator, and senior operator.

- 10 CFR 55.41 and 55.43, which provide the operator and senior operator examination requirements, respectively.

- The "Policy Statement on Engineering Expertise on Shift," published in the *Federal Register* (50 FR 43621) on October 28, 1985, which gives licensees the option to combine the senior operator and shift technical advisor into a "dual role" position.

- NUREG-0800, "Standard Review Plan," Section 13.1.2-13.1.3, "Operating Organization," Acceptance Criterion C.1, which requires that a shift supervisor with a senior operator's license, who is also a member of the station supervisory staff, be on site at all times when at least one unit is loaded with fuel.

- NUREG-0800, "Standard Review Plan," Section 13.1.2-13.1.3 "Operating Organization," Acceptance Criterion C.2, which requires that an auxiliary operator (nonlicensed) be assigned to the control room when a reactor is operating.

- NUREG-0800, "Standard Review Plan," Section 13.1.2-13.1.3 "Operating Organization," Acceptance Criterion C.6, which requires that, "Assignment, stationing, and relief of operators and senior operators within the control room shall be as described in Regulatory Guide 1.114."

- NUREG-0800, "Standard Review Plan," Section 13.1.2-13.1.3 "Operating Organization," Acceptance Criterion D, which requires that staffing plans include total complements of licensed personnel of no less than that required for five shift rotations.

- NUREG/CR-6838, "Technical Basis for Regulatory Guidance for Assessing Exemption Requests from the Nuclear Power Plant Licensed Operator Staffing Requirements Specified in 10 CFR 50.54(m)," which provides the technical basis for the guidance presented in this document.

2. REVIEW OF THE CONCEPT OF OPERATIONS

2.1 Discussion

The purpose of reviewing the concept of operations is to provide the reviewer with a more comprehensive understanding of how the proposed staffing and associated exemption requests fit into the overall design and operation of the plant. At the most general level, the term, *concept of operations* refers to a description of how the design, systems, and operational characteristics of a plant, such as an advanced reactor, relate to a licensee's or applicant's organizational structure, staffing, and management framework.

Concept of operations may also be used when discussing a system. For example, an applicant may intend to add an intelligent monitoring system that will monitor plant parameters and take control actions that were previously performed by an operator in response to certain conditions. The concept of operations in this case would describe the purpose of the new system, its relationship to other systems, the system's characteristics and operations, and user interactions with the system, as well as training and procedures requirements.

2.2 Applicant Submittals

The applicant should submit a concise, but complete, description of how the plant or system design, operation, and management necessitate the exemption request and where supporting information is available in the applicant's overall submittal. Also, any new or redefined terms should be discussed as they apply within the concept of operations. The concept of operations should describe the following elements:

- the primary design and operating characteristics of the plant or system and the specific staffing goals and assumptions necessary to implement the concept of operations

- the number of personnel who will have plant monitoring and operational control responsibilities on each shift (i.e., "control personnel") and staffing levels for these personnel across shifts

- the roles and responsibilities of each individual designated as control personnel, if that individual is responsible for control and monitoring plant or unit operations

- the training and qualifications required for control personnel.

- the overall operating environment and primary HSIs to be used by control personnel

- the interaction of control personnel with automated systems, including responsibilities for monitoring, operating, and overriding automated systems

- the interaction of control personnel with automated support systems and the role of these systems in the overall management and control of the plant

- other mechanisms that enable or support control personnel responsibilities for monitoring, disturbance detection, situation assessment, response planning, response execution, and the management of transitions between automatic and manual control

- the interactions of control personnel with each other and with people not directly responsible for the control and safe operation of the plant

- multi-unit operations

- modular unit operations

- operations during construction of additional units

2.3   Review Criteria

The reviewer should confirm that the applicant's description of the concept of operations for the plant or system is complete and that the applicant has addressed each of the aspects of operations and roles of the control personnel.

2.4   Additional Resources

- AIAA G-043-1992: *Guide for the Preparation of Operational Concept Documents*, (American Institute of Aeronautics and Astronautics, 1993).

- NUREG-0711: *Human Factors Engineering Program Review Model*, Section 8.4.2 Concept of Operations, (NRC, 2004).

3. REVIEW THE OPERATIONAL CONDITIONS

3.1 Discussion

The purpose of the review of the operational conditions is to ensure that the operational conditions which present the greatest potential challenges to the effective and safe performance of control personnel, under the conditions of the requested exemption, were analyzed by the applicant and support the exemption request. During the normal course of the licensing process, an applicant is required to analyze the full range of operational conditions that personnel will be required to manage. For the purpose of justifying a given exemption or set of exemptions to 10 CFR 50.54(m), however, it may be unnecessary to analyze the full range of potential conditions.

NUREG-0711, Section 11.4.1, "Operational Conditions Sampling," provides a robust set of guidelines for identifying operational conditions for use in the verification and validation of control room designs. This same guidance can be used for the purpose of reviewing requests for exemptions from 10 CFR 50.54(m). The focus of the sampling should be adjusted to the conditions requested in the exemption and the emphasis should be on those operational conditions known to present the greatest challenges to human performance.

The exemption request should provide a discussion of the rationale for selecting specific conditions and for not analyzing others. The reviewer should assess whether or not this set of operations is reasonable, based on the design of the plant, the concept of operations, and the range of operational conditions that could be considered. The reviewer should request additional clarification or justification if the applicant's set of operational conditions is incomplete.

3.2 Applicant Submittals

At a minimum, this section of the exemption request should include the following elements:

* a description of the operational conditions selected for analysis

* the rationale for selecting the operational conditions analyzed and for excluding others that could have been analyzed

3.3 Review Criteria

The reviewer should be able to ensure that the applicable criteria described in the following sections have been met.

3.3.1 Operational Conditions Sampling for an Advanced Reactor Design

The reviewer should confirm that the following operational conditions were analyzed or that an adequate rationale for not analyzing the conditions was provided:

- normal operational events, including plant startup, shutdown, or refueling, and significant changes in operating power

- failure events, including instrument failures and HSI failures

- transients and accidents

- reasonable, risk-significant, and beyond-design-basis events, derived from the plant-specific probabilistic risk assessment (PRA)

- conditions that challenge plant safety functions as a result of interconnections and interactions among systems

The reviewer should confirm that the following types of personnel tasks were included in the analysis:

- risk-significant human actions

- difficult tasks identified through the operating experience review

- a range of procedure-guided tasks that are well defined by normal, abnormal, emergency, alarm response, and test procedures

- a range of knowledge-based tasks that require greater reasoning about safety and operating goals and the various means of achieving them

- a range of human cognitive activities, including decision-making

- a range of human interactions, including tasks performed by individual control personnel and any tasks performed by personnel acting as a crew

- tasks that are performed with high frequency

- tasks that are important or difficult, but infrequently performed

The reviewer should confirm that the analysis included the following situational factors that are known to challenge human performance:

- operationally difficult tasks

- error-forcing contexts

- high-workload conditions

- varying-workload situations

- fatigue and circadian factors

- environmental factors

Finally, the reviewer should confirm that the range and combination of operational conditions considered by the applicant are appropriate and adequate.

3.3.2 Special Considerations for Plant Modification Programs

The reviewer should confirm that the following criteria have been fulfilled:

- the operational conditions selected include the tasks that are affected by the modification, rather than the entire range of tasks required to analyze a plant design

- transfer of learning effects on human performance were assessed when a new system is replacing an existing HSI, when procedures have been modified, or when personnel will be required to use both the new system and an existing HSI

- the potential for deactivated HSIs that will be left in place to interfere with task performance was considered

- the range and combination of operational conditions considered by the applicant were appropriate and adequate

3.4 Additional Resources

- NUREG-0711: *Human Factors Engineering Program Review Model*, Section 11.4.1 Operational Conditions Sampling, (NRC, 2004).

- NUREG-1513: *Integrated Safety Analysis Guidance Document*, (NRC, 2001).

- NUREG/CR-6393: *Integrated System Validation: Methodology and Review Criteria*, (O'Hara et al., 1997).

4. REVIEW OPERATING EXPERIENCE

4.1   Discussion

The purpose of this step of the review is to ensure that the applicant has performed a review of relevant operating experience to identify and address staffing-related lessons learned that may be important to the exemption request. The purpose of the applicant's review of operating experience should be to identify previous staffing-related problems in order to avoid repeating them, should the exemption request be approved. It is also used for identifying similar staffing practices that have proven to be effective and successful implementations of similar technologies and concepts of operation.

The amount of relevant operating experience available will vary, depending upon whether the exemption request involves new reactor designs or the introduction of new systems into an existing plant. The greatest amount of information will be available for systems and staffing practices that have been implemented in other nuclear power plants. Information regarding new system designs and staffing practices should also be sought from other industries in which similar systems or practices have been implemented (e.g., chemical manufacturing plants, other types of power generating plants, some military systems).

The results of the applicant's operating experience review may be used as input to several of the exemption request analyses. For example, the applicant's operating experience review may identify problematic operations and tasks that should be considered in the selection of operational conditions and tasks to be analyzed. Experience regarding the impacts of staffing shortfalls may also be useful in the task analysis, for defining jobs, and in developing the staffing plan. Effective implementations of technologies may be used as the basis for allocating functions between the technologies and control personnel. Operating experience may also provide data to support the staffing plan verification and validation process.

4.2   Applicant Submittals

Operating experience may be available from the following sources:

*   predecessor plants or systems

*   plants or systems using similar technologies, practices, or concepts of operation

*   recognized industry human performance and staffing issues

*   issues identified by predecessor or similar plant personnel

*   prototype or experimental plants/systems

*   experience from other industries

4.3   Review Criteria

The reviewer should be able to ensure that each of the criteria below has been met, as applicable:

- Predecessor or similar plants and systems included in the analysis are identified and their similarities and differences from the exemption under consideration are described.

- Any recognized industry issues with the plant or system design are identified.

- Any recognized industry issues with staffing for similar plants, systems, or technologies are identified.

- Other sources of operating experience data are identified, along with any limitations of their use in performing the review for the exemption requested.

- The applicant has reviewed the staffing goals and numbers of control personnel for each of the related plants or systems selected.

- The process used by the applicant for identifying issues during the operating experience review includes a description of the assumptions, criteria, and constraints used in selecting issues and developing interviews of control personnel.

- The applicant has identified the risk-important actions associated with existing plants, systems, or relevant technologies that could potentially be a problem if the requested exemption is granted.

- The operating experience review was of sufficient scope to identify the most important relevant information and that the applicant's rationale for excluding some experience that could have been analyzed is reasonable.

Examples of effective implementations of technologies, practices, or concepts of operation included as support for the exemption are fully substantiated and documented.

4.4   Additional Resources

- IAEA Safety Series No. I 75-INSAG-3: *Basic Safety Principles for Nuclear Power Plants*, (International Atomic Energy Agency, 1988).

- IEEE Standard 845-1999 *IEEE Guide to Evaluation of Human-System Performance in Nuclear Power Generating Stations*, (Institute of Electrical and Electronics Engineers, 1999).

- NUREG-0711: *Human Factors Engineering Program Review Model*, Section 3 Operating Experience Review, (NRC, 2004).

- NUREG/CR-6400: *HFE Insights for Advanced Reactors Based Upon Operating Experience*, (Higgins and Nasta, 1996).

5. REVIEW THE FUNCTIONAL REQUIREMENTS ANALYSIS AND FUNCTION ALLOCATION

5.1 Discussion

There are two purposes for this step of the review. The first purpose is to ensure that the applicant has defined and evaluated the impact of the exemption request on the plant/system functions that must be performed to satisfy plant safety objectives. The second purpose is to ensure that the allocation of functions to humans and systems has resulted in a role for control personnel that uses human strengths, avoids human limitations, and can be performed under the operational conditions evaluated in the exemption request. A *function* is a process or activity that is required to achieve a desired goal.

Functional requirements analysis is the identification of functions that must be performed to prevent or mitigate the consequences of postulated accidents that could damage the plant or cause undue risk to the health and safety of the public. The functional requirements analysis is also conducted to identify and define functions for all other normal operating conditions with the goal of achieving effective, efficient, and safe operations. A functional requirements analysis is conducted to achieve the following objectives:

- determine the objectives, performance requirements, and constraints of the design

- define the high-level functions that have to be accomplished to meet the objectives and desired performance

- define the relationships between high-level functions and plant systems responsible for performing the function

- provide a framework for understanding the role of controllers (whether personnel or system) in controlling the plant

Function allocation is the analysis of the requirements for plant control and the assignment of control functions to personnel (e.g., manual control), system elements (e.g., automatic and passive control, self-controlling phenomena), and combinations thereof (e.g., shared control and automatic systems with manual backup).

Functional requirements and function allocation analyses are also required when implementing new systems in existing plants. Plant modifications may change the level of automation of the original design and change the functions that are allocated to systems and personnel, leading to an exemption request.

5.2 Applicant Submittals

The functional requirements analysis and function allocation data submitted in support of the exemption request should include the following:

- the set of functions identified as being relevant to the exemption request

- the sequence of performance of the functions, triggering events for their initiation, and conditions for their completion or suspension

- minimum function performance requirements in terms of time, timing, and accuracy

- identification of functions that include risk-important human actions and the consequences (e.g., error rates or estimates of error rates) of not performing those actions, performing them incompletely, or not performing them within the time required

- a description of the allocation of functions to control personnel, automated systems, or a combination of the two

- a description of how the allocation of functions supports integrated control staff roles across functions and systems

- a description of how control personnel functions relate to the functions performed by other plant personnel

- identification of functions that can be reallocated across or between control personnel, automated systems, or other plant staff, and a description of the strategies and criteria employed for reallocation

- identification of functions with risk-important human actions that may be reallocated with a description of how the risks are managed through the reallocation

- identification of function allocations that may affect the roles, responsibilities, or qualifications for licensed control personnel

- identification of function allocations to any new control personnel jobs

- applicable supporting data from the concept of operations, the operational conditions defined, and the operating experience review

5.3 Review Criteria

The reviewer should be able to ensure that each of the criteria below has been met:

- The set of functions identified as applicable to the analysis is complete and appropriately characterized.

- All functions have been allocated to control personnel, automated systems, or a combination of the two, and that the strategies and criteria for the allocations are clear and met.

- The function allocations support integrated control staff roles across functions, systems, and other plant personnel.

- Any new or modified licensed control personnel positions resulting from the function requirements analysis and function allocation have been identified and characterized.

- The data analyses were performed using appropriate parameters and methods.

- The assumptions and estimates used in conducting the analyses were documented and appropriate.

5.4   Additional Resources

- IAEA-TECDOC-668: *The Role of Automation and Humans in Nuclear Power Plants*, (International Atomic Energy Agency, 1992).

- IEEE Std. 1023-1988: *IEEE Guide to the Application of Human Factors Engineering to Systems, Equipment, and Facilities of Nuclear Power Generating Stations*, (Institute of Electrical and Electronics Engineers, 1988).

- NUREG-0711: *Human Factors Engineering Program Review Model*, Section 4 Functional Requirements Analysis and Function Allocation, (NRC, 2004).

- NUREG/CR-3331: *A Methodology for Allocation of Nuclear Power Plant Control Functions to Human and Automated Control*, (Pulliam et al., 1983).

6. REVIEW THE TASK ANALYSIS

6.1 Discussion

The purpose of the task analysis review is to ensure that the applicant's analysis identifies the specific tasks that are needed to accomplish functions and their staffing implications. The functions allocated to plant personnel define their jobs. Human actions are performed to accomplish these functions. Human actions can be further divided into tasks. A *task* is a group of related activities that have a common objective or goal. *Task analysis* is the identification of requirements for accomplishing these tasks (i.e., for specifying the requirements for the displays, data process, controls, and job aids needed to accomplish tasks).

The scope of the task analyses performed by an applicant will vary, depending upon the nature of the design or system(s) for which the exemption request has been initiated. In the case of a modification to an existing plant, the task analysis should address the tasks that have changed. In the case of a new plant control room design, the task analysis should address the set of tasks that control personnel will be required to perform for the defined operational conditions.

For each task, the information, control, and task support requirements should be addressed by the applicant's task analysis, as applicable. The information should be used to identify issues of task timing, workload, and situation awareness and to determine resource conflicts that would affect staffing assignments.

A number of acceptable methods exist for conducting task analyses. The reviewer should ensure that the applicant has used a generally recognized acceptable approach.

6.2 Applicant Submittals

The task analysis data submitted in support of the exemption request should include the following, as applicable:

- the set of tasks identified as being relevant to the exemption request

- the sequence of performance of the tasks, triggering events for their initiation, and conditions for their completion or suspension

- minimum task performance requirements in terms of time, timing, accuracy, or other relevant criteria, as identified in Table 2

- identification of tasks that include risk-important human actions and the consequences (e.g., error rates or estimates of error rates) of not performing those actions, performing them incompletely, or not performing them within the time required

- identification of tasks that may affect the roles, responsibilities, or qualifications for licensed control personnel

- identification of tasks for any new control personnel jobs

- applicable supporting data from the concept of operations, the operational conditions defined, function requirements analysis and function allocation, and the operating experience review

Table 2. Task Performance Requirements

Category	Data Item	Requirements
Information Requirements	Alarms and alerts	Any alarms and alerts that would trigger a task to start
	Parameters	Any parameters that would indicate the task is appropriate for performance
	Feedback needed to indicate adequacy of actions taken	Any parameter that the operator would need to monitor during the task to ensure the task is correctly executed
Decision making Requirements	Decision type (relative, absolute, probabilistic)	Explanation of how and when decisions between alternative tasks are made
	Evaluations to be performed	Parameters that must be evaluated in the decision and how they are applied
	Coordination	Decisions that must be made or approved by others
Response Requirements	Type of action to be taken	A description of the operator action taken in the task

Table 2 Task Performance Requirements (Con't)

Category	Data Item	Requirements
Response Requirements (Con't)	Task frequency	A measurement of how frequently the task occurs
	Task tolerance	A measure of the allowable accuracy for the task to be considered successfully performed
	Task accuracy	The expected value of how accurately the task will be performed by the operator
	Consequences of inaccurate performance	The effect that inaccurate task performance has on other tasks in the scenario
	Time available and temporal constraints	The time allowable for the operator to complete the task
	Time required	An estimate of the amount of time required for the operator to complete the task. Statistical distributions should be provided. If distributions are unavailable, a typical minimum and maximum time should be provided.
	Physical position	The physical position and location required for the operator to perform the task
	Biomechanics	A description of the physical activity that must be performed (movements) and the forces required
Communication Requirements	Personnel communication for monitoring or control, including among control personnel and directing the activities of others	A description of the participants in the communication and information communicated
	Personnel communication for administrative, reporting, and external communications	A description of the participants in the communication and information communicated
Workload	Visual	A ranking of the visual workload
	Auditory	A ranking of the auditory workload

Table 2 Task Performance Requirements (Con't)

Category	Data Item	Requirements
Workload (Con't)	Cognitive	A ranking of the cognitive workload
	Psychomotor	A ranking of the psychomotor workload
	Overlap of task requirements	An indicator if other tasks may or may not be run in parallel with this task
Task Support Requirements	Special protective clothing	Any clothing that could interfere with task performance or be required for task performance
	Job aids or reference materials needed	Any reference materials that could improve performance, or be required to perform the task, and any demands for multiple, concurrent use
	Tools and equipment needed	Any tools or equipment required to perform the task
	Automation or automated support	Any automated support systems that could affect performance or be required to perform the task, and any demands for multiple, concurrent use
Workplace Factors	Ingress and egress paths to work site	Any specific paths an operator must take to get to the work area
	Workspace envelope needed by action taken	Any space requirements needed to perform the task
	Typical and extreme environmental conditions	Measures of the typical and extreme conditions for— • lighting • heat • temperature • noise

Table 2 Task Performance Requirements (Con't)

Category	Data Item	Requirement
Situational and Performance Shaping Factors	Stress	Level of stress expected based upon the severity of the scenario or conditions
	Reduced staffing	Reasonable expectations about understaffing in the scenario
	Fatigue	Typical and extreme conditions for— • time since last sleep • point in circadian cycle
Hazard Identification	Identification of hazards involved	Any hazards that may impair performance or make an operator unavailable due to injury

(Adapted from NUREG-0711, Table 5.1, "Task Considerations")

6.3   Review Criteria

The reviewer should be able to ensure that each of the following criteria has been met:

•   The set of tasks identified as applicable to the analysis is complete and appropriately characterized.

•   The task performance requirements for each task were comprehensively identified.

•   The tasks for any new or modified licensed control personnel positions (as specified in 10 CFR Part 55) have been identified and characterized.

•   The data analyses were performed using appropriate parameters and methods.

•   The assumptions and estimates used in conducting the analyses were documented and appropriate.

6.4   Additional Resources

•   IEC 964: *Design for Control Rooms of Nuclear Power Plants*, (International Electrotechnical Commission, 1989).

- IEEE Std. 1023-1988: *IEEE Guide to the Application of Human Factors Engineering to Systems, Equipment, and Facilities of Nuclear Power Generating Stations*, (Institute of Electrical and Electronics Engineers, 1988).

- NUREG-0711: *Human Factors Engineering Program Review Model*, Section 5 Task Analysis, page 21, (NRC, 2004).

- NUREG/CR-3371: *Task Analysis of Nuclear Power Plant Control Room Crews*, (Burgey et al., 1983).

- NUREG/CR-6690: *The Effects of Interface Management Tasks on Crew Performance and Safety in Complex, Computer-Based Systems*, (O'Hara and Brown, 2002).

7. REVIEW THE JOB DEFINITIONS

7.1   Discussion

The purpose of the job definition review is to confirm that the applicant has established clear and rational job definitions for the personnel who will be responsible for controlling the plant. For an existing plant in which new systems will be implemented, the purpose of the review is to ensure that the applicant has retained clear and rational job definitions for control room personnel. A *job* is defined as the group of tasks and functions that are assigned to a personnel position. A *job definition* specifies the responsibilities, authorities, knowledge, skills, and abilities that are required to perform the tasks and functions assigned to a job.

An applicant's job definitions should describe the impact of the exemption request on each job affected. For example, an exemption request could entail redefining and reassigning the functions and tasks of the current senior operator position. Current senior operator responsibilities for coordinating and overseeing the activities of reactor operators in a control room located onsite could be eliminated, partially reallocated to intelligent monitoring systems, and/or assigned to off-site personnel who monitor on-site activities remotely.

Alternatively, a new job could be created that has no analogue in an existing plant or under the current regulations. As a hypothetical example, a specialist job could be created in which an individual is uniquely trained and qualified to troubleshoot the software that supports new systems or new HSIs, and to assume control if systems fail and backups must be used.

A job that consists of interrelated responsibilities and authorities that do not conflict would be coherent. A classic example of conflicting responsibilities would be a Senior Operator in a traditional control room, who is charged with maintaining an overview of operational conditions. These additional responsibilities may compromise his or her ability to maintain "the big picture." Conflicting responsibilities, in the past, have included responsibilities for taking control actions or responding to information requests from personnel outside of the control room. The reviewer should ensure that the applicant's job definitions appropriately prioritize the responsibilities of each position and do not incorporate role conflicts.

An important aspect of the job definition review is to ensure that the qualifications required for each position are delineated. The qualifications required for a plant staff position consist of the knowledge, skills, and abilities/aptitudes (KSAs) an individual must possess to meet the performance criteria established for the tasks assigned to the position. The information derived from the function and task analyses should provide a basis for identifying the required KSAs for each position.

The job definition review will be necessary for all exemption requests. Its scope should be limited to the jobs of licensed personnel that are impacted by the exemption request. Within a job, the scope of the review may also be limited by the extent (e.g., only a few job functions or tasks impacted) and character (e.g., only responsibilities affected, not qualifications) of the exemption request.

7.2   Applicant Submittals

The job definition data submitted in support of the exemption request should include the following:

- a description of the scope and the impacts of the exemption request on the roles, responsibilities, and qualifications of control personnel

- identification of any new or modified roles, responsibilities, and qualifications for licensed control room personnel (under the current requirements) included in the exemption request

- identification of the roles, responsibilities, and qualifications for any new jobs included in the exemption request

- applicable data from the concept of operations, operational conditions, operating experience, functional requirements analysis and function allocation, and task analysis for each of the jobs affected that support the roles and responsibilities identified in the exemption request

- applicable data from the KSA analysis for each of the jobs affected that support the qualifications identified in the exemption request

- a final job description for each job impacted by the exemption request

- job definitions which appropriately prioritize the responsibilities of each position and that do not incorporate role conflicts

7.3   Review Criteria

The reviewer should be able to ensure that each of the following criteria has been met:

- The scope and impact of the exemption request is adequately addressed for control personnel jobs.

- Applicable data from the concept of operations, operational conditions, operating experience, functional requirements analysis and function allocation, and task analysis support the roles and responsibilities assigned to each impacted job in the exemption request.

- The KSA analysis is complete, and the KSAs are consistent with the qualifications required for each impacted job identified in the exemption request.

- Coherent job descriptions are maintained for licensed control room personnel (under the current requirements), or are defined for any new jobs included as a part of the exemption request.

- The job definitions for control personnel who will work in crews are coordinated.

7.4   Additional Resources

- Information Notice 93-81: *Implementation of Engineering Expertise on Shift*, (NRC, 1981).

- Generic Letter 86-04: *Policy Statement on Engineering Expertise on Shift*, (NRC, 1986).

- NUREG-0711: *Human Factors Engineering Program Review Model*, Section 6 Staffing and Qualifications, (NRC, 2004).

- NUREG-1122: *Knowledge and Abilities Catalog for Nuclear Power Plant Operators: Pressurized Water Reactors*, (NRC 1998).

- NUREG-1123: *Knowledge and Abilities Catalog for Nuclear Power Plant Operators: Boiling Water Reactors*, (NRC, 1998).

- NUREG-1220: *Training Review Criteria and Procedures*, (NRC, 1993).

- Regulatory Guide 1.149: *Nuclear Power Plant Simulation Facilities for Use in Operator License Examinations*, (NRC, 1996).

- Regulatory Guide 1.8: *Qualification and Training of Personnel for Nuclear Power Plants*, (NRC, 2000).

- Regulatory Guide 1.114: *Guidance to Operators and to Senior Operators in the Control Room of a Nuclear Power Plant*, (NRC, 1989).

- *Code of Federal Regulations, Title 10, "Energy,"* Part 50.120, "Training and qualification of nuclear power plant personnel."

8. REVIEW THE STAFFING PLAN

8.1 Discussion

The purpose of the staffing plan review is to ensure that the applicant has systematically analyzed the requirements for the numbers of qualified personnel that are necessary to operate the plant safely under the operational conditions analyzed. That is, the staffing plan should answer the question, "How many individuals must be qualified and available to fill each job?"

The applicant's staffing plan should be supported by the results of the functional requirements analysis and function allocation, task analyses, and the job definitions for each position required under the operational conditions considered. In addition, the applicant's submittal should define the proposed shift composition and shift scheduling. *Shift composition* refers to the different types of jobs that must be filled on each shift and the number of personnel required for each of the jobs on a shift. In the case of remote operations or operations that will take place outside of a traditional control room, the applicant should also define the locations of the personnel comprising a shift.

8.2 Applicant Submittals

The staffing plan submitted in support of the exemption request should include the following elements:

- the set of operational conditions considered for the staffing plan, to the extent that they differ from those submitted for other elements of the exemption request

- the proposed staffing levels, shift composition, and shift schedules for the identified operational conditions

- a description of how the staffing plan supports integrated staff roles across shifts and operational conditions

- identification of the types of control personnel who can be substituted within each job, given the concept of operations

- expected travel times or response times for control personnel who need to move to new locations (e.g., home to the plant or office) or provide other support (e.g., to log in to system control computers from home), when applicable

- a description of how control personnel staffing relates to the larger plant staffing and the support roles that control personnel may play in the larger staffing context

- applicable supporting data from the concept of operations, the set of operational conditions considered, the functional requirements analysis and function allocation, task analysis, job definitions, and the operating experience review

8.3 Review Criteria

The reviewer should be able to ensure that each of the following criteria has been met:

- The set of operational conditions identified as applicable to the staffing plan is complete and representative of the exemption request.

- The staffing plan will provide an adequate number of qualified personnel to operate the plant safely under the operational conditions considered.

- Roles and responsibilities are integrated across shifts and among personnel.

- Travel and response times are adequate and do not trigger adverse conditions for the safety of the plant.

- The staffing plan uses data from previous sections in a logical/rational manner.

8.4   Additional Resources

- ANSI/ANS 3.1: *Selection, Qualification, and Training of Personnel for Nuclear Power Plants,* (American Nuclear Society, 1993).

- ANSI/ANS 58.8: *Time Response Design Criteria for Safety-Related Operator Actions,* (American National Standards Institute, 1994).

- Generic Letter 86-04: *Policy Statement on Engineering Expertise on Shift,* (NRC, 1986).

- Information Notice 93-81: *Implementation of Engineering Expertise on Shift,* (NRC, 1981).

- Information Notice 95-48: *Results of Shift Staffing Study,* (NRC, 1995).

- Information Notice 97-78: *Crediting of Operator Actions in Place of Automatic Actions and Modifications of Operator Actions, Including Response Times,* (NRC, 1997).

- NUREG-0711: *Human Factors Engineering Program Review Model,* Section 6 Staffing and Qualifications, (NRC, 2004).

- NUREG/IA-0137: *A Study of Control Room Staffing Levels for Advanced Reactors,* (Hallburt and Morisseau, 2000).

- Regulatory Guide 1.8: *Personnel Selection and Training,* (NRC, 2000).

- Regulatory Guide 1.114: *Guidance to Operators at the controls and to Senior Operators in the Control Room of a Nuclear Power Unit,* (NRC, 1986).

9. REVIEW OF ADDITIONAL DATA AND ANALYSES

Applicants may provide additional supporting data and analyses as part of the exemption request submittals. These additional materials should be reviewed based on their applicability to the requested exemption and the need for the supporting data and analyses. Additional review may include the following areas:

- human reliability analysis used to demonstrate the impacts of risk-important human actions

- human-system integration data used to demonstrate that the design of the HSIs supports the concept of operations, functional requirements analysis and function allocation, task analysis, staffing plan, and operating experience

- KSA analysis used in support of new or changing job definitions

- KSA analysis used to support modified tasks or human-system interfaces

- procedures and training documentation used to demonstrate the implementation of components of the concept of operations, functional requirements analysis and function allocation, or task analysis

The reviewer should also consider additional submittals that would be expected, based on the character of the exemption request. For example, if remote support operations are proposed, data supporting the new communications skills required for control personnel may be appropriate. NUREG-0711 includes review criteria for these areas.

10. REVIEW THE STAFFING PLAN VALIDATION

10.1   Discussion

The purpose of reviewing the validation of the staffing plan is to ensure that the applicant fully considered the dynamic interactions between the plant design, its systems, and control personnel for the operational conditions identified for the exemption request. *Staffing plan validation* refers to an evaluation using performance-based tests to determine whether the staffing plan meets performance requirements and acceptably supports safe operation of the plant.

The applicant should provide data or demonstrations that the control personnel specified in the staffing plan can satisfy the plant and human performance requirements identified in the functional requirements analysis, function allocation, and task analyses. These data or demonstrations may come from operating experience, human-in-the-loop simulations, human performance models, or a mix of these methods. The data or demonstrations should include the full range of operational conditions identified for the exemption request, as well as a reasonable representation of the human performance variability expected in the context of the operational conditions.

10.1.1 Operational Conditions Sampling

The applicant should include the operational conditions relevant to the exemption request in the staffing plan validation. As a practical matter, however, it may be unnecessary to address all of the possible variations of these conditions. It may be reasonable to combine some of them into scenarios.

The applicant's submittal should identify the operational conditions included in each scenario. The submittal should identify the key plant and system parameters relevant to the scenario and the state of these parameters at the start of the scenario, during critical transition points in the scenario, at times when action by control personnel is expected, the results of control actions, and the status of the parameters at the end of the scenario. The submittal should also identify the criteria for determining successful performance of the plant, system, and control personnel within the scenarios. The submittal should sample a sufficient number of operational conditions such that the personnel and plant performance are challenged.

10.1.2 Human Performance Measures and Criteria

This section discusses "how" the data may have been collected. The reviewer needs to be aware of the methods and conditions under which data were collected to be able to assess the analyses. The applicant needs to identify the measures of human performance used to evaluate individual and crew performance of the control personnel in the scenarios. Outcome-oriented human performance measures include the following examples (among others):

- time to complete actions

- timeliness of actions

- accuracy and completeness of actions

- omitted actions

Outcome measures can usually be observed, measured directly, and be linked to overall plant and system performance measures. The measures may be aggregated to the crew level in evaluating crew performance, and ultimately, the adequacy of the staffing plan.

Measures of conditions that can affect the response of control personnel should also be addressed. To the extent that environmental conditions such as heat, cold, or lighting need to be considered, their impacts on control personnel performance should be addressed. If shift durations or scheduling have the potential to cause sleep loss and fatigue among control personnel, these impacts will need to be assessed as well. The impacts of these types of conditions are most often seen as degradations in control personnel performance, which may not always result in degraded system performance or failure. Degraded personnel performance increases the risk of failure, however, so the frequency and extent to which control personnel are exposed to the adverse conditions should be assessed.

Time and information processing demands placed on the control personnel may also degrade performance. The impacts of these types of demands can be assessed using measures of cognitive workload and situation awareness.

Cognitive workload refers to the degree to which an individual's cognitive and perceptual capabilities are taxed during the performance of the tasks that comprise his or her job. Most cognitive workload measures are structured self-reports from the users of a system regarding the time pressure they experience, the mental effort involved in performing their tasks, and the amount of stress they experience. Excessive cognitive workload will lead to performance decrements, such as delays, inaccurate responses, errors in diagnoses, and omissions.

Situation or *situational awareness* (SA) is defined as an individual's mental model of what has happened, the current status of the system, and what will happen in the next brief time period (Endsley & Garland, 2000). Because the quality of a person's decision selection and performance is determined by the "goodness" (accuracy, completeness, relevance) of the internal, or mental, model of the system, it is critical for control personnel to form and maintain complete and accurate SA. To determine if new plant designs and/or new staffing arrangements adequately support SA, it is important to evaluate the degree to which all control personnel demonstrate adequate SA.

In addition to defining the measures of human performance used in validating the staffing plan, the applicant should identify the criteria established to determine the acceptability of the results obtained. The criteria may include the following examples:

- Nominal task performance times will not be exceeded by more than 10 percent.
- No more than 60 seconds will be required to begin Task X after Event Y.
- Temperatures will be maintained within ± 5 degrees.
- No actions will be omitted.

Other performance measures and criteria may be derived from relevant requirements in Table 2 of Section 6 of this document.

10.1.3 Data Sources or Demonstration Methods

The data sources or demonstration methods used by the applicant to validate the staffing plan may include operating experience, human-in-the-loop simulations, human performance models, or a mix of these methods. Although the applicant may submit other types of data or demonstrations, the reviewer should ensure that the submittals assess the dynamic interactions between the plant design, its systems, and control personnel for the operational conditions identified for the exemption request.

Key considerations for the data sources or methods include the following aspects:

- the range of operational conditions considered

- how well the behavior of the plant or systems is represented

- how well the behavior of control personnel, including a reasonable representation of the human performance variability that may be expected, is represented

The first two considerations are straightforward to assess. However, a reasonable representation of human performance variability requires some explanation. First, human performance is variable, both within individuals and across individuals. Within the context of high reliability systems, such as nuclear power plants, this variability is generally sufficient to require representation in the analysis of human performance. Therefore, the reviewer needs to assure that the applicant considered human performance variability and describe how it was addressed in the validation process.

The applicant will select among the possible data sources and methods for validating the staffing plan based upon the availability, quality, and comprehensiveness of the data from each of the possible sources. The reviewer should ensure that the applicant has developed and executed a plan for integrating these data and for demonstrating that the dynamic interactions between the plant design, its systems and control personnel for the operational conditions defined for the exemption request have been fully considered. Finally, the applicant will need to report on the outcomes of the staffing validation in a way that clearly demonstrates to the reviewer that the staffing plan supports the requested exemption.

10.1.3.1 Data from Operating Experience

Data from operating experience tend to carry high face validity. They are most useful when they are drawn from similar plants, technologies, or organizations that are implementing similar concepts of operation. The longer the duration of successful operation or success in mitigating unwanted events, the more support operating experience can provide to the staffing plan and exemption request. Data from training or licensing of control personnel that demonstrate effective performance may also be considered, particularly for operational conditions that have never actually occurred or that have occurred at low frequencies.

10.1.3.2 Data from Human-in-the-Loop Simulations

Using human-in-the-loop simulations for staffing plan validation will often be limited by simulator availability. Simulators may be difficult to access for validation purposes since they are often in heavy use for training or licensing examinations. Further, for new or modified plants or systems, there may be no, or a limited number of, control personnel who have the qualifications and capabilities to perform the roles of the "humans" in the loop. These factors may limit the range of human performance issues that can be assessed.

A key benefit of data from human-in-the-loop simulations is that they can represent a wide range of operational conditions, often at high levels of fidelity. High-fidelity simulators are often built well in advance of the actual plants they represent, so that they may be available for use in support of an exemption request. Simulators with lower levels of fidelity may also be used to provide supporting data. For example, a simulator that reflects plant or system behavior well, but does not reflect the actual HSI, may be useful for demonstrating the time and timing of events and available control personnel response times. In either case, these techniques can be used to capture quantitative, objective measures and criteria to support the exemption request.

10.1.3.3 Data from Human Performance Models

Human performance models typically require data such as tasks, task times and timing, flow logics, and error probabilities. Although these data are typically available from the task analysis, some of the data frequently need to be estimated. The models also require algorithms to represent performance variations, as well as measures of factors such as cognitive workload. Because the models are projections of human performance that are often based on a limited amount of concrete data, they are subject to challenge. These limitations are moderated by the fact that (1) there are representations for which validation exists, and (2) the models can be exercised across a range of values for critical parameters to assess the model's sensitivity.

Data from human performance models can provide a robust representation of the performance of control personnel across the range of operational conditions. Models can easily incorporate the various conditions that may affect human performance, human performance variability, and measures of concepts, such as cognitive workload and situation awareness. Although human performance models historically have incorporated plant or system representations of limited fidelity, human performance models can now be linked to more sophisticated plant or system simulations. The human performance models also make it relatively easy to assess different staffing alternatives. As with the human-in-the-loop simulations, quantitative, objective measures and criteria can be captured to support the exemption request.

10.2   Applicant Submittals

The reviewer should confirm that data for the following four areas are provided, as applicable.

10.2.1 Operational Conditions Sampling

The reviewer should confirm that the submittal includes the following information regarding operational conditions sampling:

- a description of each of the scenarios used in validating the staffing plan

- a description of how the scenarios incorporate the operational conditions relevant to the exemption request

- a description of system and key plant parameters relevant to the scenarios

- relevant criteria for evaluating successful performance

- scenarios that challenge personnel, plant, and system performance

10.2.2 Human Performance Measures and Criteria

The reviewer should confirm that the submittal includes the following information regarding human performance measures and criteria:

- a listing of the human performance measures and criteria identified for the validation and a discussion of the rationale for their inclusion, as well as for the exclusion of other reasonable measures for the individual and the crew

- descriptions of relationships for those measures and criteria specific to the data sources or methods used or whose definitions vary across the methods

- identification, description, and definition of any measures and criteria specific to methods or constructs (e.g., cognitive workload or situation awareness measurement)

- descriptions of environmental or external influences that could impact human performance and how they are integrated into the assessment

- time and information processing standards and how they are incorporated into the assessment

- the type of data source

10.2.3 Data Sources or Demonstration Methods

The reviewer should confirm that the submittal includes the following information regarding data sources or demonstration methods:

*   a description of the integrated design and execution of the validation using the selected sources and methods, validation method, or implementation plan description

*   a description of the data sources and methods used, the parts of the validation each supports, and how they have been integrated

*   a description of limitations in the scope and data quality (e.g., plant/system similarities and differences, assumptions, estimates, algorithms, numbers and qualifications of subjects) for each source

*   a description of how dynamic interactions were assessed

10.2.4 Staffing Plan Validation Outcomes

The reviewer should confirm that the submittal includes the following information regarding the outcomes of the staffing plan validation:

*   a description and analysis of the outcomes of the staffing plan validation

*   workload demands

*   situational awareness

10.3   Review Criteria

The reviewer should be able to ensure that the applicable criteria described in the following sections have been met:

10.3.1 Operational Conditions Sampling

The reviewer should confirm that the following criteria have been met, as applicable:

*   The scenarios fully incorporated the operational conditions relevant to the exemption request.

*   Relevant criteria were used to evaluate successful performance.

*   Scenarios relevant to the exemption request were used.

*   Scenarios that challenge the personnel, plant, and system were used.

10.3.2 Human Performance Measures and Criteria

The reviewer should confirm that the following criteria have been met, as applicable:

- The human performance measures and criteria are relevant to the plant/system concept of operations.

- At a minimum, the selected human performance measures represent the most important outcome behaviors.

- The rationale for excluding some potential human performance measures is reasonable.

- The selected measures assess both individual and crew performance, where appropriate.

- Measures specific to data collection methods or constructs have been used appropriately.

- The criteria defined for acceptable human performance on each measure are reasonable.

- Any identified environmental conditions, external conditions, or staffing practices that could potentially degrade individual or crew performance, are effectively addressed by the staffing plan.

- Valid methods and criteria have been identified.

- The data analyses were performed using appropriate parameters and methods.

- The assumptions and estimates used in conducting the analyses are documented and appropriate.

10.3.3 Data Sources and Demonstration Methods

The reviewer should confirm that the following criteria have been met, as applicable:

* The selected design of the staffing plan validation, the data sources, and the demonstration methods comprehensively address the dynamic aspects of the staffing plan and support the requested exemption.

* The data sources and demonstration methods were used appropriately.

* The appropriate quantitative, objective measures and criteria were defined and captured. See Section 5.2, "Validating Staffing Plans," of NUREG/CR-6838 for further information.

* The data collection and analysis were conducted appropriately.

* The scope and data quality were adequate.

* The outcomes were reasonable and valid.

10.3.4 Staffing Plan Validation Outcomes

The reviewer should confirm that the following criteria have been met, as applicable:

* The results of analyses demonstrate that control personnel, individually and working in crews, if applicable, can accomplish their tasks within performance criteria.

* The results of analyses demonstrate that the staffing plan does not result in either excessively high or minimal workload demands on control personnel for the operational conditions considered.

* The results of the analyses demonstrate that the staffing plan does not compromise control personnel situational awareness.

* The staffing plan effectively addressed any identified environmental conditions or staffing practices that could potentially degrade individual or crew performance.

10.4 Additional Resources

- ANSI/AIAA G-035-1992: *Guide to Human Performance Measurements*, (American National Standards Institute, 1993).

- IEEE Std. 845-1999: IEEE *Guide to the Evaluation of Human-System Performance in Nuclear Power Generating Stations*, (Institute of Electrical and Electronics Engineers, 1999).

- NUREG-0711: *Human Factors Engineering Program Review Model*, Section 11.4.3 Integrated System Validation, (NRC, 2004).

- NUREG/CR-6393: *Integrated System Validation: Methodology and Review Criteria*, (O'Hara et al., 1997).

- NUREG/CR-6838: *Technical Basis for Regulatory Guidance for Assessing Exemption Requests from the Nuclear Power Plant Licensed Operator Staffing Requirements Specified in 10 CFR 50.54(m)*, (Plott, C., T. Engh, and V. Barnes, 2003).

11. DETERMINE THE ACCEPTABILITY OF THE EXEMPTION REQUEST

In this step, NRC staff must make a final decision regarding the acceptability of the exemption request. The decision will be based on the aggregate findings from the previous steps of the review. The reviewer should be able to satisfactorily answer the following questions regarding the acceptability of the exemption request:

- Was sufficient justification provided to ensure that the impacts of the exemption request were adequately addressed in the following components:
  - concept of operations
  - operational conditions
  - operating experience
  - functional requirements analyses and function allocation (or reallocation)
  - task analyses
  - job definitions
  - staffing plan
  - additional supporting data and analyses
  - verification and validation of the staffing plan

- Were the range and combination of operational conditions considered by the applicant appropriate and adequate?

- Were the data analyses performed using appropriate parameters and methods?

- Were the assumptions and estimates used in conducting the analyses documented and appropriate?

- Will acceptance of the exemption request provide at least the same level of assurance that public health and safety are maintained as the current regulations require?

The reviewer should prepare a summary of the overall findings along with the determination of the acceptability of the exemption request. If the reviewer determines that there is insufficient evidence to support the exemption request, the reviewer should identify the limitations of the submittals and the further analyses, data, or changes in the exemption request that are needed. The reviewer should generate an RAI and/or develop a letter indicating the weaknesses and strengths of the exemption request.

APPENDIX A

REVIEW CHECKLISTS

STEP 1. REVIEW THE EXEMPTION REQUEST

Purpose: The review of the exemption request is performed to ensure that the reviewer understands the scope of the request and to ensure that the applicant has submitted the necessary information to perform the review.

Applicability: This section is applicable in all cases for which an exemption from 10 CFR 50.54(m) is requested.

Instructions: Verify that the request includes the following information. If the information is not provided or deemed not applicable, please indicate the reason in the comments field.

Y	N	N/A	Data and information contain:	Comments
			Description of the specific aspects of 10 CFR 50.54(m) from which an exemption is requested.	
			Physical representation of the plant and systems involved.	
			Descriptions of plant/system responses to inputs and expected response times	
			Detailed representation of the control room, control suites, and/or the HSI to be used for monitoring and control actions.	
			Definitions of any new terms used or redefinitions of terms whose meanings are changed.	
			Information to meet the data requirements of subsequent review steps.	

Instructions: Confirm that each of the following review criteria has been met. If a criterion has not been met or has been deemed not applicable, please indicate the reason in the comments field.

Y	N	N/A	Review Criteria	Comments
			One or more exemptions to 10 CFR 50.54(m) is required.	
			Exemptions from other, related regulations are either unnecessary or have been appropriately identified and described by the applicant	
			The terms used in the submittal are fully defined.	
			Adequate data and information have been submitted to meet the data requirements for the remainder of the review.	

Step 2. REVIEW OF THE CONCEPT OF OPERATIONS

Purpose: The purpose of reviewing the concept of operations is to provide the reviewer with a more comprehensive understanding of how the proposed staffing and associated exemption requests fit into the overall design and operation of the plant.

Applicability: This section is applicable in all cases for which an exemption from 10 CFR 50.54(m) is requested.

Instructions: Verify that the request includes the following information. If the information is not provided or deemed not applicable, please indicate the reason in the comments field.

Y	N	N/A	Data and information contain:	Comments
			The primary design and operating characteristics of the plant or system and the specific staffing goals and assumptions necessary to implement the concept of operations.	
			The number of personnel who will have plant monitoring and operational control responsibilities on each shift (i.e., "control personnel"), and staffing levels for these personnel across shifts.	
			The roles and responsibilities of each individual designated as a control personnel are provided, if that individual is responsible for control and monitoring plant or unit operations.	
			The training and qualifications required for control personnel.	
			The overall operating environment and primary HSI to be used by control personnel.	
			The interaction of control personnel with automated systems including responsibilities for monitoring, operating, and overriding automated systems.	

Y	N	N/A	Data and information contain (con't):	Comments
			The interaction of control personnel with automated support systems and the role of these systems in the overall management and control of the plant.	
			Other mechanisms that enable or support control personnel responsibilities for monitoring, disturbance detection, situation assessment, response planning, response execution, and the management of transitions between automatic and manual control.	
			The interactions of control personnel with each other and with people not directly responsible for the control and safe operation of the plant.	
			Multi-unit operations.	
			Modular unit operations	
			Operation during construction of additional units.	

Instructions: Confirm that each of the following review criteria has been met. If a criterion has not been met or has been deemed not applicable, please indicate the reason in the comments field.

Y	N	N/A	Review Criteria	Comments
			The applicant's description of the concept of operations for the plant or system is complete.	
			Each of the aspects of operations and roles of the control personnel are addressed.	

Step 3. REVIEW THE OPERATIONAL CONDITIONS

Purpose: The purposed of the review of the operational conditions is to ensure that the operational conditions which present the greatest potential challenges to the effective and safe performance of control personnel, under the conditions of the requested exemption, were analyzed by the applicant and support the exemption request.

Applicability: This section is applicable in all cases for which an exemption fro 10 CFR 50.54(m) is requested. However, the review criteria do vary depending upon whether the exemption request is for an advanced reactor control room or for a modification to an existing plant control room.

Instructions: Verify that the request includes the following information. If the information is not provided or deemed not applicable, please indicate the reason in the comments field.

Y	N	N/A	Data and information contain:	Comments
			A description of the operational conditions selected for analysis.	
			The rationale for selecting the operational conditions analyzed and for excluding others that could have been analyzed.	

Instructions: Confirm that each of the following review criteria has been met. If a criterion has not been met or has been deemed not applicable, please indicate the reason in the comments field.

Use the following criteria for an advanced reactor control room exemption request review:

Y	N	N/A	Review Criteria	Comments
			Normal operational events, including plant startup, shutdown, or refueling, and significant changes in operating power were analyzed.	
			Failure events, including instrument failures and HSI failures, were analyzed.	

Y	N	N/A	Review Criteria (Con't)	Comments
			Transients and accidents were analyzed	
			Reasonable, risk-significant, and beyond-design-basis events, derived from the plant-specific PRA, were analyzed.	
			Conditions that challenge plant safety functions as a result of interconnections and interactions among systems were analyzed.	
			Risk-significant human actions were included in the analysis.	
			Difficult tasks identified through the operating experience review were included in the analysis.	
			A range of procedure-guided tasks that are well defined by normal, abnormal, emergency, alarm response, and test procedures were included in the analysis.	
			A range of knowledge-based tasks that require greater reasoning about safety and operating goals and the various means of achieving them were included in the analysis.	
			A range of human cognitive activities, including decision making were included in the analysis.	
			A range of human interactions, including tasks performed by individual control personnel and any tasks performed by personnel acting as a crew were included in the analysis.	
			Tasks that are performed with high frequency were included in the analysis.	
			Tasks that are important or difficult but infrequently performed were included in the analysis.	

Y	N	N/A	Review Criteria (Con't)	Comments
			Operationally difficult tasks were included in the analysis.	
			Error-forcing contexts were included in the analysis.	
			High-workload conditions were included in the analysis.	
			Varying-workload situations were included in the analysis.	
			Fatigue and circadian factors were included in the analysis.	
			Environmental factors were included in the analysis.	
			The range and combination of operational conditions considered by the applicant are appropriate and adequate.	

Use the following criteria for modification to an existing plant control room exemption request review:

Y	N	N/A	Review Criteria	Comments
			The operational conditions selected include the tasks that are affected by the modification, rather than the entire range of tasks required to analyze a plant design.	
			Transfer of learning effects on human performance were assessed when a new system is replacing an existing HSI, when procedures have been modified, or when personnel will be required to use both the new system and an existing HSI.	
			The potential for deactivated HSI that will be left in place to interfere with task performance was considered.	
			The range and combination of operational conditions considered by the applicant were appropriate and adequate.	

STEP 4. REVIEW OPERATING EXPERIENCE

Purpose: The purpose of this step of the review is to ensure that the applicant has performed a review of relevant operational experience to identify and address staffing-related lessons learned that may be important to the exemption request.

Applicability: This section is applicable in all cases for which and exemption from 10 CFR 50.54(m) is requested. The amount of relevant experience available will vary, depending upon whether the exemption request involves new reactor designs or the introduction of new systems into an existing plant.

Instructions: Verify that the request includes the following information. If the information is not provided or deemed not applicable, please indicate the reason in the comments field.

Y	N	N/A	Data and information contain:	Comments
			Operating experience from predecessor plants or systems.	
			Operating experience from plants or systems using similar technologies, practices, or concepts of operation.	
			Recognized industry human performance and staffing issues.	
			Issues identified by predecessor or similar plant personnel.	
			Operating experience from prototype or experimental plants/systems.	
			Operating experience from other industries.	

Instructions: Confirm that each of the following review criteria has been met. If a criterion has not been met or has been deemed not applicable, please indicate the reason in the comments field.

Y	N	N/A	Review Criteria	Comments
			Predecessor or similar plants and systems included in the analysis are identified and their similarities and differences from the exemption under consideration are described.	
			Any recognized industry issues with the plant or system design are identified.	
			Any recognized industry issues with staffing for similar plants, systems, or technologies are identified.	
			Other sources of operating experience data are identified, along with any limitations of their use in performing the review for the exemption requested.	
			For each of the related plants or systems selected, the applicant has reviewed the staffing goals and numbers of control personnel.	
			The process used by the applicant for identifying issues during the operating experience review includes a description of the assumptions, criteria, and constraints used in selecting issues and developing interviews of control personnel.	
			The applicant has identified the risk-important actions associated with existing plants, systems, or relevant technologies that could potentially be a problem if the requested exemption is granted.	

Y	N	N/A	Review Criteria (Con't)	Comments
			The operating experience review was of sufficient scope to identify the most important relevant information and the applicant's rationale for excluding some experience that could have been analyzed is reasonable.	
			Examples of effective implementations of technologies, practices, or concepts of operation included as support for the exemption are fully substantiated and documented.	

STEP 5. REVIEW FUNCTIONAL REQUIREMENTS ANALYSIS AND FUNCTION ALLOCATION

Purpose: The first purpose of this step of the review is to ensure that the applicant has defined and evaluated the impact of the exemption request on the plant/system functions that must be performed to satisfy plant safety objectives. The second purpose is to ensure that the allocation of functions to humans and systems has resulted in a role for control personnel that uses human strengths, avoids human limitation, and can be performed under the operational conditions evaluated in the exemption request.

Applicability: This section is applicable in all cases for which an exemption from 10 CFR 50.54(m) is requested.

Instructions: Verify that the request includes the following information. If the information is not provided or deemed not applicable, please indicate the reason in the comments field.

Y	N	N/A	Data and information contain:	Comments
			The set of functions identified as being relevant to the exemption request.	
			The sequence of performance of the functions, triggering events for their initiation, and conditions for their completion or suspension.	
			Minimum function performance requirements in terms of time, timing, and accuracy.	
			Identification of functions that include risk-important human actions and the consequences (e.g., error rates or estimates of error rates) of not performing those actions, performing them incompletely, or not performing them within the time required.	
			A description of the allocation of functions to control personnel, automated systems, or a combination of the two.	
			A description of how the allocation of functions supports integrated control staff roles across functions and systems.	

Y	N	N/A	Data and information contain (Con't):	Comments
			A description of how control personnel functions relate to the functions performed by other plant personnel.	
			Identification of functions that can be reallocated across or between control personnel, automated systems, or other plant staff, and a description of the strategies and criteria employees for reallocation.	
			Identification of functions with risk-important human actions that may be reallocated with a description of how the risks are managed through the reallocation.	
			Identification of function allocations that may affect the roles, responsibilities, or qualifications for licensed control personnel.	
			Identification of function allocations to any new control personnel jobs.	
			Applicable supporting data from the concept of operations, the operational conditions defined, and the operating experience review.	

Instructions: Confirm that each of the following review criteria has been met. If a criterion has not been met or has been deemed not applicable, please indicate the reason in the comments field.

Y	N	N/A	Review Criteria	Comments
			The set of functions identified as applicable to the analysis is complete and appropriately characterized.	
			All functions have been allocated to control personnel, automated systems, or a combination of the two, and that the strategies and criteria for the allocations are clear and met.	

Y	N	N/A	Review Criteria (Con't)	Comments
			The function allocations support integrated control staff roles across functions, systems, and other plant personnel.	
			Any new or modified licensed control personnel positions resulting from the function requirements analysis and function allocation have been identified and characterized.	
			The data analyses were performed using appropriate parameters and methods.	
			The assumptions and estimates used in conducting the analyses were documented and appropriate.	

STEP 6. REVIEW THE TASK ANALYSIS

Purpose: The purpose of the task analysis review is to ensure that the applicant's analysis identifies the specific tasks that are needed to accomplish functions and their staffing implications.

Applicability: This section is applicable in all cases for which an exemption from 10 CFR 50.54(m) is requested.

Instructions: Verify that the request includes the following information. If the information is not provided or deemed not applicable, please indicate the reason in the comments field.

Y	N	N/A	Data and information contain:	Comments
			The set of tasks identified as being relevant to the exemption request.	
			The sequence of performance of the tasks, triggering events for their initiation, and conditions for their completion or suspension.	
			Minimum task performance requirements in terms of time, timing, accuracy, or other relevant criteria, as identified in the table of task considerations that follow.	
			Identification of tasks that include risk-important human actions and the consequences (e.g., error rates or estimates of error rates) of not performing those actions, performing them incompletely, or not performing them within the time required.	
			Identification of tasks that may affect the roles, responsibilities, or qualifications for licensed control personnel.	
			Identification of tasks for any new control personnel jobs.	
			Applicable supporting data from the concept of operations, the operational conditions defined, function requirements analysis and function allocation, and the operating experience review.	

Table of Task Considerations

Y	N	N/A	Category	Data Item	Requirements	Comments
			Information Requirements	Alarms and alerts	Any alarms and alerts that would trigger a task to start.	
				Parameters	Any parameters that would indicate the task is appropriate for performance.	
				Feedback needed to indicate adequacy of actions taken	Any parameter that the operator would need to monitor during the task to ensure the task is correctly executed.	
			Decision making Requirements	Decision type (relative, absolute, probabilistic)	Explanation of how and when decisions between alternative tasks are made.	
				Evaluations to be performed	Parameters that must be evaluated in the decision and how they are applied.	
				Coordination	Decisions that must be made or approved by others.	
			Response Requirements	Type of action to be taken	A description of the operator action taken in the task.	
				Task frequency	A measurement of how frequently the task occurs.	
				Task tolerance	A measure of the allowable accuracy for the task to be considered successfully performed.	
				Task accuracy	The expected value of how accurately the task will be performed by the operator.	

Table of Task Considerations (Con't)

Y	N	N/A	Category	Data Item	Requirements	Comments
			Response Requirements (Con't)	Consequences of inaccurate performance	The effect that inaccurate task performance has on other tasks in the scenario.	
				Time required	An estimate of the amount of time required for the operator to complete the task. Statistical distributions should be provided. If distributions are unavailable, a typical minimum and maximum time should be provided.	
				Physical position	The physical position and location required for the operator to perform the task.	
				Biomechanics	A description of the physical activity that must be performed (movements) and the forces required.	
			Communication Requirements	Personnel communication for monitoring or control, including among control personnel and directing the activities of others	A description of the participants in the communication and information communicated.	
				Personnel communication for administrative, reporting, and external communications	A description of the participants in the communication and information communicated.	
			Workload	Visual	A ranking of the visual workload.	
				Auditory	A ranking of the auditory workload	
				Cognitive	A ranking of the cognitive workload	

Table of Task Considerations (Con't)

Y	N	N/A	Category	Data Item	Requirements	Comments
			Workload (con't)	Psychomotor	A ranking of the psychomotor workload	
				Overlap of task requirements	An indicator if other tasks may or may not be run in parallel with this task.	
			Task Support Requirements	Special protective clothing	Any clothing that could interfere with task performance or be required for task performance.	
				Job aids or reference materials needed	Any reference materials that could improve performance, or be required to perform the task, and any demands for multiple, concurrent use	
				Tools and equipment needed	Any tools or equipment required to perform the task	
				Automation or automated support	Any automated support systems that could affect performance or be required to perform the task, and any demands for multiple, concurrent use	
			Workplace Factors	Ingress and egress paths to work site	Any specific paths an operator must take to get to the work area	
				Workspace envelope needed by action taken	Any space requirements needed to perform the task	

Table of Task Considerations (Con't)

Y	N	N/A	Category	Data Item	Requirements	Comments
			Workplace Factors (con't)	Typical and extreme environmental conditions	Measures of the typical and extreme conditions for— • lighting • heat • temperature • noise	
			Situational and Performance Shaping Factors	Stress	Level of stress expected based upon the severity of the scenario or conditions	
				Reduced staffing	Reasonable expectations about understaffing in the scenario	
				Fatigue	Typical and extreme conditions for— • time since last sleep • point in circadian cycle	
			Hazard Identification	Identification of hazards involved	Any hazards that may impair performance or make an operator unavailable due to injury	

Instructions: Confirm that each of the following review criteria has been met. If a criterion has not been met or has been deemed not applicable, please indicate the reason in the comments field.

Y	N	N/A	Review Criteria	Comments
			The set of tasks identified as applicable to the analysis is complete and appropriately characterized.	
			The task performance requirements for each task were comprehensively identified.	
			The tasks for any new or modified licensed control personnel positions have been identified and characterized.	
			The data analyses were performed using appropriate parameters and methods.	
			The assumptions and estimates used in conducting the analyses were documented and appropriate.	

STEP 7. REVIEW THE JOB DEFINITIONS

Purpose: The purpose of the job definition review is to confirm that the applicant has established clear and rational job definitions for the personnel who will be responsible for controlling the plant. For an existing plant in which new systems will be implemented, the purpose of the review is to ensure that the applicant has retained clear and rational job definitions for control room personnel.

Applicability: This section is applicable in all cases for which an exemption from 10 CFR 50.54(m) is requested. Its scope should be limited to the jobs of control room personnel that are impacted by the exemption request. Within a job, the scope of the review may also be limited by the extent (e.g., only a few job functions or tasks impacted) and character (e.g., only responsibilities affected, not qualifications) of the exemption request.

Instructions: Verify that the request includes the following information. If the information is not provided or deemed not applicable, please indicate the reason in the comments field.

Y	N	N/A	Data and information contain:	Comments
			A description of the scope and the impacts of the exemption request on the roles, responsibilities, and qualifications of control personnel.	
			Identification of any new or modified roles, responsibilities, and qualifications for licensed control room personnel (under the current requirements) included in the exemption request.	
			Identification of the roles, responsibilities, and qualifications for any new jobs included in the exemption request.	
			Applicable data from the concept of operations, operational conditions, operating experience, functional requirements analysis and function allocation, and task analysis for each of the jobs affected that support the roles and responsibilities identified in the exemption request.	
			Applicable data from the KSA analysis for each of the jobs affected that support the qualifications identified in the exemption request.	

Y	N	N/A	Comments	Data and information contain (Con't):
				A final job description for each job impacted by the exemption request.
				Job definitions which appropriately prioritize the responsibilities of each position and that do not incorporate role conflicts.

Instructions: Confirm that each of the following review criteria has been met. If a criterion has not been met or has been deemed not applicable, please indicate the reason in the comments field.

Y	N	N/A	Comments	Review Criteria
				The scope and impact of the exemption request on control personnel jobs.
				Applicable data from the concept of operations, operational conditions, operating experience, functional requirements analysis and function allocation, and task analysis support the roles and responsibilities assigned to each impacted job in the exemption request.
				The KSA analysis is complete and that the KSAs are consistent with the qualifications required for each impacted job identified in the exemption request.
				Coherent job descriptions are maintained for licensed control room personnel (under the current requirements), or are defined for any new jobs included as a part of the exemption request
				The job definitions for control personnel who will work in crews are coordinated.

STEP 8. REVIEW THE STAFFING PLAN

Purpose: The purpose of the staffing plan review is to ensure that the applicant has systematically analyzed the requirements for the numbers of qualified personnel that are necessary to operate the plant safely under the operational conditions analyzed.

Applicability: This section is applicable in all cases where an exemption from 10 CFR 50.54(m) is requested.

Instructions: Verify that the request includes the following information. If the information is not provided or deemed not applicable, please indicate the reason in the comments field.

Y	N	N/A	Data and information contain:	Comments
			The set of operational conditions considered for the staffing plan, to the extent that they differ from those submitted for other elements of the exemption request.	
			The proposed staffing levels, shift composition, and shift schedules for the identified operational conditions.	
			A description of how the staffing plan supports integrated staff roles across shifts and operational conditions.	
			Identification of the types of control personnel who can be substituted within each job, given the concept of operations.	
			When applicable, expected travel times or response times for control personnel who need to move to new locations (e.g., home to the plant or office) or provide other support (e.g., to log in to system control computers from home).	

Y	N	N/A	Data and information contain (Con't):	Comments
			A description of how control personnel staffing relates to the larger plant staffing and the support roles that control personnel may play in the larger staffing context.	
			Applicable supporting data from the concept of operations, the set of operational conditions considered, the functional requirements analysis and function allocation, task analysis, job definitions, and the operating experience review.	

Instructions: Confirm that each of the following review criteria has been met. If a criterion has not been met or has been deemed not applicable, please indicate the reason in the comments field.

Y	N	N/A	Review Criteria	Comments
			The set of operational conditions identified as applicable to the staffing plan is complete and representative of the exemption request.	
			The staffing plan will provide adequate numbers of qualified personnel to operate the plant safely under the operational conditions considered.	
			Roles/responsibilities are integrated across shifts and among personnel.	
			Travel and response times are adequate and do not trigger adverse conditions for the safety of the plant.	
			The staffing plan uses data from previous sections in a logical/rational manner.	

STEP 9. REVIEW ADDITIONAL DATA AND ANALYSES

Purpose: The purpose of the review of additional data and analyses is to allow the consideration of additional data that is often not applicable, but in some cases may be applicable, in the review of the exemption request.

Applicability: This section is applicable in all cases for which an exemption from 10 CFR 50.54(m) is requested. However, in most cases these data and analyses are unnecessary for the evaluation of staffing. The reviewer must determine if any additional analyses are needed based upon the specific exemption request presented.

Instructions: Determine if each of the following additional analyses are necessary. If the analysis is needed, please indicate the reason in the comments field.

Y	N	N/A	Additional Analysis	Comments
			Human reliability analysis used to demonstrate the impacts of risk-important human actions.	
			Human-system integration data used to demonstrate that the design of the HSI supports the concept of operations, functional requirements analysis and function allocation, task analysis, staffing plan, and operating experience.	
			Knowledge, skills, and abilities analysis used in support of new or changing job definitions.	
			Knowledge, skills, and abilities analysis used to support modified tasks or HSI.	
			Procedures and training documentation used to demonstrate the implementation of components of the concept of operations, functional requirements analysis and function allocation, or task analysis.	

STEP 10. REVIEW THE STAFFING PLAN VALIDATION

Purpose: The purpose of reviewing the validation of the staffing plan is to ensure that the applicant fully considered the dynamic interactions between the plant design, its systems, and control personnel for the operational conditions identified for the exemption request.

Applicability: This section is applicable in all case for which an exemption from 10 CFR 50.54(m) is requested.

Instructions: Verify that the request includes the following information. If the information is not provided or deemed not applicable, please indicate the reason in the comments field.

Operational Conditions Sampling

Y	N	N/A	Data and information contain:	Comments
			A description of each of the scenarios used in validating the staffing plan.	
			A description of how the scenarios incorporate the operational conditions relevant to the exemption request.	
			A description of system and key plant parameters relevant to the scenarios.	
			Relevant criteria for evaluating successful performance.	
			Scenarios that challenge personnel, plant, and system performance.	

Human Performance Measures and Criteria

Y	N	N/A	Data and information contain:	Comments
			A listing of the human performance measures and criteria identified for the validation and a discussion of the rationale for their inclusion, as well as for the exclusion of other reasonable measures for the individual and the crew.	
			Descriptions of relationships for those measures and criteria specific to the data sources or methods used or whose definitions vary across the methods.	
			Identification, description, and definition of any measures and criteria specific to methods or constructs (e.g., cognitive workload or situation awareness measurement).	
			Descriptions of environmental or external influences that could impact human performance and how hey are integrated into the assessment.	
			Time and information processing standards and how they are incorporated into the assessment.	
			The type of data source.	

Data Sources or Demonstration Methods

Y	N	N/A	Data and information contain:	Comments
			A description of the integrated design and execution of the validation using the selected sources and methods, validation method, or implementation plan description.	
			A description of the data sources and methods used, the parts of the validation each supports, and how they have been integrated.	

Y	N	N/A	Data and information contain (Con't):	Comments
			A description of limitations in the scope and data quality (e.g., plant/system similarities/differences, assumptions, estimates, algorithms, numbers/qualifications of subjects) for each source.	
			A description of how dynamic interactions were assessed.	

Staffing Plan Validation Outcomes

Y	N	N/A	Data and information contain:	Comments
			A description and analysis of the outcomes of the staffing plan validation.	
			Workload demands.	
			Situational awareness.	

Instructions: Confirm that each of the following review criteria has been met. If a criterion has not been met or has been deemed not applicable, please indicate the reason in the comments field.

Operational Conditions Sampling

Y	N	N/A	Review Criteria	Comments
			The scenarios fully incorporate the operational conditions relevant to the exemption request.	
			Relevant criteria for evaluation of successful performance were used.	
			Scenarios relevant to the exemption request were used.	
			Scenarios that challenge the personnel, plant, and system were used.	

Human Performance Measures and Criteria

Y	N	N/A	Review Criteria	Comments
			Confirm that the human performance measures and criteria are relevant to the plant/system concept of operations.	
			The human performance measures selected, at a minimum, represent the most important outcome behaviors.	
			The rationale for excluding some potential human performance measures is reasonable.	
			The measures selected assess both individual and crew performance, where appropriate.	
			The criteria defined for acceptable human performance on each measure is reasonable.	
			Any identified environmental conditions, external conditions, or staffing practices that could potentially degrade individual or crew performance, are effectively addressed by the staffing plan.	
			Valid methods and criteria have been identified.	
			The data analyses were performed using appropriate parameters and methods.	
			The assumptions and estimates used in conducting the analyses are documented and appropriate.	

Data Sources or Demonstration Methods

Y	N	N/A	Review Criteria	Comments
			The design of the staffing plan validation, the data sources, and the demonstration methods selected comprehensively address the dynamic aspects of the staffing plan and support the requested exemption.	
			The data sources and demonstration methods were used appropriately.	
			The data collection and analysis were conducted appropriately.	
			The scope and data quality were adequate	
			The outcomes were reasonable/valid.	

Staffing Plan Validation Outcomes

Y	N	N/A	Review Criteria	Comments
			The results of analyses demonstrate that control personnel, individually and working in crews, if applicable, can accomplish their tasks within performance criteria.	
			Results of analyses demonstrate that the staffing plan does not result in either excessively high or minimal workload demands on control personnel for the operational conditions considered.	
			The results of the analyses demonstrate that the staffing plan does not compromise control personnel situational awareness.	
			Any identified environmental conditions or staffing practices that could potentially degrade individual or crew performance are effectively addressed by the staffing plan.	

STEP 11. DETERMINE THE ACCEPTABILITY OF THE EXEMPTION REQUEST

Purpose: The purpose of this step is to make a final decision regarding the acceptability of the exemption request.

Applicability: This section is applicable in all cases for which an exemption from 10 CFR 50.54(m) is requested.

Instructions: Confirm that each of the following review criteria has been met. If a criterion has not been met or has been deemed not applicable, please indicate the reason in the comments field.

Y	N	N/A	Review Criteria	Comments
			Was sufficient justification provided to ensure that the impacts of the exemption request were adequately addressed in the concept of operations documentation?	
			Was sufficient justification provided to ensure that the impacts of the exemption request were adequately addressed in the operational conditions documentation?	
			Was sufficient justification provided to ensure that the impacts of the exemption request were adequately addressed in the operating experience documentation?	
			Was sufficient justification provided to ensure that the impacts of the exemption request were adequately addressed in the functional requirements analyses and function allocation documentation?	
			Was sufficient justification provided to ensure that the impacts of the exemption request were adequately addressed in the task analyses documentation?	
			Was sufficient justification provided to ensure that the impacts of the exemption request were adequately addressed in the job definitions documentation?	

Y	N	N/A	Review Criteria (Con't)	Comments
			Was sufficient justification provided to ensure that the impacts of the exemption request were adequately addressed in the staffing plan documentation?	
			Was sufficient justification provided to ensure that the impacts of the exemption request were adequately addressed in the additional supporting data and analyses documentation?	
			Was sufficient justification provided to ensure that the impacts of the exemption request were adequately addressed in the verification and validation of the staffing plan documentation?	
			Were the range and combination of operational conditions considered by the applicant appropriate and adequate?	
			Were the data analyses performed using appropriate parameters and methods?	
			Were the assumptions and estimates used in conducting the analyses documented and appropriate?	
			Will acceptance of the exemption request provide at least the same level of assurance that public health and safety are maintained as the current regulations require?	

APPENDIX B

GLOSSARY

GLOSSARY

10 CFR 50.54, Conditions of licenses - The conditions that must be met in a nuclear power plant in order for a license to be issued.

10 CFR 50.54(m) - The minimum shift staffing requirements that must currently be met for a license to be issued for a nuclear power plant.

Advanced control room - A control room that is primarily based on digital technology. It typically provides the primary operator interaction with the plant via computer-based interfaces, such as video display units. This is in contrast to "conventional" control rooms, which provide the primary operator interaction with the plant via analog interfaces, such as gauges.

Advanced reactor - A nuclear power plant design that incorporates new technology such as advanced automation, passive safety systems, and/or new human system integration concepts.

Algorithm - A step-by-step procedure for solving a problem or accomplishing some task through a process, especially by a computer.

Cognitive workload - The degree to which a person's mental capabilities are taxed during the performance of the tasks that comprise his or her job.

Computer-supported cooperative network - The use of computers and electronic devices as a medium through which to communicate in real time

Concept of operations - A description of how the design, systems, and operational characteristics of a plant relate to an organization's structure, staffing, and management framework.

Control personnel - Individuals licensed to manipulate controls that affect the reactivity or power level of a nuclear reactor, manipulate fuel, and/or direct the activities of individuals so licensed.

Exemption application - A request for licensing that asks for an exemption from any of the requirements of 10 CFR Part 50.

Function - A process or activity that is required to achieve a desired goal.

Function allocation - The analysis of the requirements for plant control and the assignment of control functions to personnel or system elements or a combination of personnel or system elements.

Functional requirements analysis - The identification of functions that must be performed to prevent or mitigate the consequences of postulated accidents that could damage the plant or cause undue risk to the health and safety of the public.

Human reliability analysis - The process of evaluating the potential for and mechanisms of human error that may affect plant safety.

Human-system interface - The part of a system through which personnel interact to perform their functions and tasks. In this document, "system" refers to a nuclear power plant. Major HSIs include alarms, information displays, controls, and job performance aids.

Intelligent agent - Any computer system that interacts with a human to assist in cognitive processing functions or, in some cases, initiate purposeful action as a result of predictions related to the user's goal (i.e. computer-supported decision-making)

Integrated system validation - An evaluation using performance-based tests to determine whether an integrated system design (i.e., hardware, software, and personnel elements) meets performance requirements and acceptably supports safe operation of the plant.

Job - A group of tasks and functions that are assigned to a personnel position.

Job definition - The responsibilities, authorities, knowledge, skills, and abilities that are required to perform the tasks and functions assigned to a job.

Light-water reactor - A term used to describe reactors using water as coolant, including boiling-water reactors and pressurized-water reactors.

Model - A representation of how a complex entity or system is structured and functions.

Operating experience review - A review of relevant history from a plant's ongoing collection, analysis, and documentation of operating experiences; including relevant experience from other plants and/or other industries.

Passive safety feature - Design characteristics that use natural forces, such as convection and gravity, which are less dependent on active systems and components like pumps and valves to maintain plant safety.

Performance shaping factors - Factors that influence human reliability through their effects on performance, including environmental conditions, human-system interface design, procedures, training, and supervision.

Procedures - Written instructions providing guidance to plant personnel for operating and maintaining the plant and for handling disturbances and emergency conditions.

Request for exemption - An analogous term to exemption application (above).

Shift composition - The different types of jobs that must be filled on each shift and the number of personnel required for each of the jobs on a shift.

Simulator - A facility that physically represents the human-system interface configuration and that dynamically represents the operating characteristics and responses of the plant in real time.

Situation or situational awareness - An individual's mental model of what has happened, the current status of the system, and what will happen in the next brief time period.

Task - A group of related activities that have a common objective or goal.

Task analysis - The identification of requirements for accomplishing tasks (i.e., for specifying the requirements for the displays, data process, controls, and job aids needed to accomplish tasks.)

Validation - See integrated system validation (above).

Verification - The process by which the design is evaluated to determine whether it acceptably satisfies personnel task needs and human factors engineering design guidance.

Workload - The physical and cognitive demands placed on plant personnel.

APPENDIX C

REFERENCES

REFERENCES

American Institute of Aeronautics and Astronautics (1992). *Guide for the Preparation of Operational Concept Documents* (AIAA G-043-1992). Reston, Virginia: American Institute of Aeronautics and Astronautics.

American National Standards Institute/American Nuclear Society (1993). *Selection, Qualification, and Training of Personnel for Nuclear Power Plants* (ANSI/ANS 3.1). LaGrange Park, Illinois: American Nuclear Society.

American National Standards Institute (1993). *Guide to Human Performance Measurements* (ANSI/ANS 58.8). Washington, DC: American National Standards Institute.

Burgey, D., et al. (1983). *Task Analysis of Nuclear Power Plant Control Room Crews* (NUREG/CR-3371). Washington, DC: U.S. Nuclear Regulatory Commission.

Endsley, M. and D. Garland (2000). *Situation Awareness Analysis and Measurement.* Mahwah, NJ: Lawrence Earlbaum.

Hallbert, B. and D. Morisseau (2000). *A Study of Control Room Staffing Levels for Advanced Reactors* (NUREG/IA-01378). Washington, DC: U.S. Nuclear Regulatory Commission.

Higgins, J. and K. Nasta (1996). *HFE Insights for Advanced Reactors Based Upon Operating Experience* (NUREG/CR-6400). Washington, DC: U.S. Nuclear Regulatory Commission.

Institute of Electrical and Electronics Engineers (1999). *IEEE Guide to the Evaluation of Human-System Performance in Nuclear Power Generating Stations* (IEEE Std. 845-1999). New York: Institute of Electrical and Electronics Engineers.

Institute of Electrical and Electronics Engineers (1988). *IEEE Guide to the Application of Human Factors Engineering to Systems, Equipment, and Facilities of Nuclear Power Generating Stations* (IEEE Std. 1023-1988). New York: Institute of Electrical and Electronics Engineers.

International Atomic Energy Agency (1992). *The Role of Automation and Humans in Nuclear Power Plants* (IAEA-TECDOC-668). Vienna, Austria: International Atomic Energy Agency.

International Atomic Energy Agency (1988). *Basic Safety Principles for Nuclear Power Plants* (IAEA Safety Series No. I 75-INSAG-3). Vienna, Austria: International Atomic Energy Agency.

International Electrotechnical Commission (1989). *Design for Control Rooms of Nuclear Power Plants* (IEC 964). Geneva, Switzerland: Bureau Central de la Commission Electrotechnique Internationale.

NRC (2004). *Human Factors Engineering Program Review Model.* (NUREG-0711, Rev.2). Washington, DC: U.S. Nuclear Regulatory Commission.

NRC (2001). *Integrated Safety Analysis Guidance Document* (NUREG-1513). Washington, DC: U.S. Nuclear Regulatory Commission.

NRC (2000). *Qualification and Training of Personnel for Nuclear Power Plants* (Regulatory Guide 1.8, Rev. 3). Washington, DC: U.S. Nuclear Regulatory Commission.

NRC (1998). *Standard Review Plan* (NUREG-0800). Washington, DC: U.S. Nuclear Regulatory Commission.

NRC (1998). *Knowledge and Abilities Catalog for Nuclear Power Plant Operators: Boiling Water Reactors* (NUREG-1123, Rev. 2). Washington, DC: U.S. Nuclear Regulatory Commission.

NRC (1998). *Knowledge and Abilities Catalog for Nuclear Power Plant Operators: Pressurized Operator Actions, Including Response Times* (Information Notice 97-78). Washington, DC: U.S. Nuclear Regulatory Commission.

NRC (1997). *Crediting of Operator Actions in Place of Automatic Actions and Modifications of Operator Actions, Including Response Times* (Information Notice 97-78). Washington, DC: U.S. Nuclear Regulatory Commission.

NRC (1996). *Nuclear Power Plant Simulation Facilities for Use in Operator License Examinations* (Regulatory Guide 1.149, Rev. 2). Washington, DC: U.S. Nuclear Regulatory Commission.

NRC (1995). *Results of Shift Staffing Study* (Information Notice 95-48). Washington, DC: U.S. Nuclear Regulatory Commission.

NRC (1993). *Training Review Criteria and Procedures* (NUREG-1220, Rev. 1). Washington, DC: U.S. Nuclear Regulatory Commission.

NRC (1989). *Guidance to Operators and to Senior Operators in the Control Room of a Nuclear Power Plant* (Regulatory Guide 1.114). Washington, DC: U.S. Nuclear Regulatory Commission.

NRC (1986). *Policy Statement on Engineering Expertise on Shift* (Generic Letter 86-04). Washington, DC: U.S. Nuclear Regulatory Commission.

NRC (1985). "Policy Statement on Engineering Expertise on Shift." *Federal Register,* Vol. 63, No. 12: p. 43621. October 28, 1985.

NRC (1981). *Implementation of Engineering Expertise on Shift* (Information Notice 93-81). Washington, DC: U.S. Nuclear Regulatory Commission.

O'Hara, J. and W. Brown (2002). *The Effects of Interface Management Tasks on Crew Performance and Safety in Complex, Computer-Based Systems* (NUREG/CR-6690, Volumes 1 and 2). Washington, DC: U.S. Nuclear Regulatory Commission.

O'Hara, J., et al. (1997). *Integrated System Validation: Methodology and Review Criteria* (NUREG/CR-6393). Washington, DC: U.S. Nuclear Regulatory Commission.

Plott, C., T. Engh, and V. Barnes (2003). *Technical Basis for Regulatory Guidance for Assessing Exemption Requests from the Nuclear Power Plant Licensed Operator Staffing Requirements Specified in 10 CFR 50.54(m)* (NUREG/CR-6838). Washington, D.C.: U.S. Nuclear Regulatory Commission.

Pulliam, R., et al. (1983). *A Methodology for Allocation of Nuclear Power Plant Control Functions to Human and Automated Control* (NUREG/CR-3331). Washington, DC: U.S. Nuclear Regulatory Commission.

Code of Federal Regulations, Title 10, "Energy," Section 50.12, "Specific Exemptions."

Code of Federal Regulations, Title 10, "Energy," Section 50.54, "Conditions of Licenses."

Code of Federal Regulations, Title 10, "Energy," Section 50.55, "Operators' Licenses."

Code of Federal Regulations, Title 10, "Energy," Section 50.120, "Training and Qualification of Nuclear Power Plant Personnel."